what if?

what if?

Serious Scientific Answers to Absurd Hypothetical Questions

RANDALL MUNROE

JOHN MURRAY

First published in Great Britain in 2014 by John Murray (Publishers)
An Hachette UK Company

5

© xkcd Inc. 2014

A CIP catalogue record for this title is available from the British Library

Hardback ISBN 978-1-84854-957-9
Trade Paperback ISBN 978-1-84854-958-6
Ebook ISBN 978-1-84854-959-3

Book design by Christina Gleason

Lyrics from 'If I Didn't Have You' © 2011
by Tim Minchin.
Reprinted by permission of Tim Minchin.

Printed and bound by Clays Ltd, St Ives plc

John Murray policy is to use papers that are natural, renewable and recyclable products
and made from wood grown in sustainable forests. The logging and manufacturing processes
are expected to conform to the environmental regulations of the country of origin.

John Murray (Publishers)
338 Euston Road
London NW1 3BH

www.johnmurray.co.uk

QUESTIONS

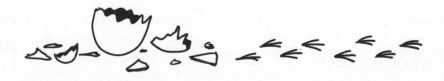

A note on units in the United Kingdom

I WAS BORN IN THE UNITED STATES, which means that as a child, I learned to think in feet, pounds, (US) gallons, and degrees Fahrenheit.

When I got a university physics degree, I had to retrain myself to use meters, kilograms, liters, and degrees Celsius.

Most people understand that metric units are in theory simpler, but until you've tried to solve a series of equations while converting back and forth between feet and miles, you don't truly appreciate the metric system.

To answer the questions in this book, I had to read a lot of scientific papers. Some of them were nice, modern articles that used metric units, but many were not. I would find myself squinting at old engineering reports—degraded to unreadability by repeated photocopying and faxing—because they contained the only known data on what happens to human skin in a supersonic wind tunnel, or the effects of nuclear weapons on filing cabinets. Those papers rarely used metric units.

I've read river engineering articles where flow rates are given in "kcfs", or kilo-cubic feet per second. I've read papers that exclusively measure volumes in acre-feet (a unit of volume equal to a foot times a chain times a furlong). I've even dealt with a modern piece of software which gives results in the ghastly "degrees Rankine" (my single least favorite unit) which is equal to degrees Fahrenheit above absolute zero.

Like the US, the UK is a land of absurd measurement systems in a world gone sane. It's true that you use more metric units than we do, but you also have things like "stone" for weight. We, in the US are confused by your different volume units;

for example, your cars get better gas mileage, since a mile per imperial gallon is better than a mile per US gallon. (To bring us closer to parity, might I suggest switching to *nautical* miles per imperial gallon?)

However, thanks to the magic of modern computing, I don't have to worry about keeping track of all these units; the computer handles it for me. I can happily enter values in furlongs and fortnights and not worry about including the conversion factors. Bit by bit, I've been letting go of my obsessive fixation on the metric system. Under the hood, all calculations are in SI units, of course, but the inputs and outputs are in whatever units I find convenient. It's freeing, and—for a physics major—a little frightening.

Now that we are at last coming free of the tyranny of manual unit conversion, we're faced with a choice. I think it's time to make up our minds: Do we want a sane, orderly system of units, or an incomprehensible patchwork of incompatible and contradictory local definitions? I vote for the second one; it sounds more entertaining.

Your imperial system is a good start. However, many of your units—such as your "feet"—are far too compatible with our US system. In light of this, I offer some modest suggestions for alternative definitions for the foot which residents of the UK could adopt:

- London Foot: Equal to 0.9977 international feet, or exactly one milliShard
- Compromise Foot: Equal to 2.14 feet, the average of one foot and one meter
- Stonehenge–Belgium Foot: Equal to 1.003 feet, or one one-millionth of the distance between Stonehenge and the edge of Belgium (which happens to be just over a million feet from Stonehenge)
- The Ice Foot: Equal to 0.7759 feet, or the diameter of a sphere of ice weighing one stone
- The Metric Foot: Equal to 10 inches
- The Metrical Foot: Equal to 10 inches followed by a stressed syllable
- The Royal Foot: Equal to the actual length of the current monarch's foot, which would be remeasured every year

We keep trying to make our units of measurement make sense. But the truth is that the world is an absurd place; why not embrace it? It's true, unit conversion errors have caused us to lose space probes once in a while. But isn't that a small price to pay for silliness?

DISCLAIMER

Do not try any of this at home. The author of this book is an Internet cartoonist, not a health or safety expert. He likes it when things catch fire or explode, which means he does not have your best interests in mind. The publisher and the author disclaim responsibility for any adverse effects resulting, directly or indirectly, from information contained in this book.

INTRODUCTION

THIS BOOK IS A collection of answers to hypothetical questions.

These questions were submitted to me through my website, where—in addition to serving as a sort of Dear Abby for mad scientists—I draw xkcd, a stick-figure webcomic.

I didn't start out making comics. I went to school for physics, and after graduating, I worked on robotics at NASA. I eventually left NASA to draw comics full-time, but my interest in science and math didn't fade. Eventually, it found a new outlet: answering the Internet's weird—and sometimes worrying—questions. This book contains a selection of my favorite answers from my website, plus a bunch of new questions answered here for the first time.

I've been using math to try to answer weird questions for as long as I can remember. When I was five years old, my mother had a conversation with me that she wrote down and saved in a photo album. When she heard I was writing this book, she found the transcript and sent it to me. Here it is, reproduced verbatim from her 25-year-old sheet of paper:

Randall:	Are there more soft things or hard things in our house?
Julie:	I don't know.
Randall:	How about in the world?
Julie:	I don't know.

Randall:	Well, each house has three or four pillows, right?
Julie:	Right.
Randall:	And each house has about 15 magnets, right?
Julie:	I guess.
Randall:	So 15 plus 3 or 4, let's say 4, is 19, right?
Julie:	Right.
Randall:	So there are probably about 3 billion soft things, and . . . 5 billion hard things. Well, which one wins?
Julie:	I guess hard things.

To this day I have no idea where I got "3 billion" and "5 billion" from. Clearly, I didn't really get how numbers worked.

My math has gotten a little better over the years, but my reason for doing math is the same as it was when I was five: I want to answer questions.

They say there are no stupid questions. That's obviously wrong; I think my question about hard and soft things, for example, is pretty stupid. But it turns out that trying to thoroughly answer a stupid question can take you to some pretty interesting places.

I still don't know whether there are more hard or soft things in the world, but I've learned a lot of other stuff along the way. What follows are my favorite parts of that journey.

RANDALL MUNROE

what if?

Q. What would happen if the Earth and all terrestrial objects suddenly stopped spinning, but the atmosphere retained its velocity?

—**Andrew Brown**

A. NEARLY EVERYONE WOULD DIE. *Then* things would get interesting.

At the equator, the Earth's surface is moving at about 470 meters per second—a little over a thousand miles per hour—relative to its axis. If the Earth stopped and the air didn't, the result would be a sudden thousand-mile-per-hour wind.

The wind would be highest at the equator, but everyone and everything living between 42 degrees north and 42 degrees south—which includes about 85 percent of the world's population—would suddenly experience supersonic winds.

The highest winds would last for only a few minutes near the surface; friction with the ground would slow them down. However, those few minutes would be long enough to reduce virtually all human structures to ruins.

TERRIBLE THINGS HAPPEN
TERRIBLE THINGS HAPPEN, BUT MORE SLOWLY

My home in Boston is far enough north to be just barely outside the supersonic wind zone, but the winds there would still be twice as strong as those in the most powerful tornadoes. Buildings, from sheds to skyscrapers, would be smashed flat, torn from their foundations, and sent tumbling across the landscape.

Winds would be lower near the poles, but no human cities are far enough from the equator to escape devastation. Longyearbyen, on the island of Svalbard in Norway—the highest-latitude city on the planet—would be devastated by winds equal to those in the planet's strongest tropical cyclones.

If you're going to wait it out, one of the best places to do it might be Helsinki, Finland. While its high latitude—above 60°N—wouldn't be enough to keep it from being scoured clean by the winds, the bedrock below Helsinki contains a sophisticated network of tunnels, along with a subterranean shopping mall, hockey rink, swimming complex, and more.

No buildings would be safe; even structures strong enough to survive the winds would be in trouble. As comedian Ron White said about hurricanes, "It's not *that* the wind is blowing, it's *what* the wind is blowing."

Say you're in a massive bunker made out of some material that can withstand thousand-mile-per-hour winds.

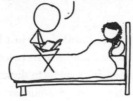

That's good, and you'd be fine . . . if you were the only one with a bunker. Unfortunately, you probably have neighbors, and if the neighbor upwind of you has a less-well-anchored bunker, your bunker will have to withstand a thousand-mile-per-hour impact by *their* bunker.

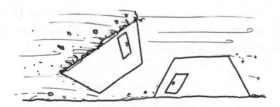

The human race wouldn't go extinct.[1] In general, very few people above the surface would survive; the flying debris would pulverize anything that wasn't nuclear-hardened. However, a lot of people below the surface of the ground would survive just fine. If you were in a deep basement (or, better yet, a subway tunnel) when it happened, you would stand a good chance of surviving.

There would be other lucky survivors. The dozens of scientists and staff at the Amundsen–Scott research station at the South Pole would be safe from the

---

[1] I mean, not right away.

winds. For them, the first sign of trouble would be that the outside world had suddenly gone silent.

The mysterious silence would probably distract them for a while, but eventually someone would notice something even stranger:

THE SUN ISN'T MOVING.

OH, THE EARTH MUST HAVE STOPPED SPINNING, DESTROYING EVERYTHING IN A GLOBAL TEMPEST.

I *HATE* IT WHEN THAT HAPPENS.

I'LL KICK IT AND SEE IF IT STARTS AGAIN.

The air

As the surface winds died down, things would get weirder.

The wind blast would translate to a heat blast. Normally, the kinetic energy of rushing wind is small enough to be negligible, but this would not be normal wind. As it tumbled to a turbulent stop, the air would heat up.

Over land, this would lead to scorching temperature increases and—in areas where the air is moist—global thunderstorms.

At the same time, wind sweeping over the oceans would churn up and atomize the surface layer of the water. For a while, the ocean would cease to have a surface at all; it would be impossible to tell where the spray ended and the sea began.

Oceans are *cold*. Below the thin surface layer, they're a fairly uniform 4°C. The tempest would churn up cold water from the depths. The influx of cold spray into superheated air would create a type of weather never before seen on Earth—a roiling mix of wind, spray, fog, and rapid temperature changes.

This upwelling would lead to blooms of life, as fresh nutrients flooded the upper layers. At the same time, it would lead to huge die-offs of fish, crabs, sea turtles, and animals unable to cope with the influx of low-oxygen water from the depths. Any animal that needs to breathe—such as whales and dolphins—would be hard-pressed to survive in the turbulent sea-air interface.

The waves would sweep around the globe, east to west, and every east-facing shore would encounter the largest storm surge in world history. A blinding cloud of sea spray would sweep inland, and behind it, a turbulent, roiling wall of water would advance like a tsunami. In some places, the waves would reach many miles inland.

The windstorms would inject huge amounts of dust and debris into the atmosphere. At the same time, a dense blanket of fog would form over the cold ocean surfaces. Normally, this would cause global temperatures to plummet. And they would.

At least, on one side of the Earth.

If the Earth stopped spinning, the normal cycle of day and night would end. The Sun wouldn't completely stop moving across the sky, but instead of rising and setting once a day, it would rise and set once a *year*.

Day and night would each be six months long, even at the equator. On the day side, the surface would bake under the constant sunlight, while on the night side the temperature would plummet. Convection on the day side would lead to massive storms in the area directly beneath the Sun.[2]

In some ways, this Earth would resemble one of the tidally locked exoplanets commonly found in a red dwarf star's habitable zone, but a better comparison might be a very early Venus. Due to its rotation, Venus—like our stopped Earth—keeps the same face pointed toward the Sun for months at a time. However, its thick atmosphere circulates quite quickly, which results in the day and the night side having about the same temperature.

Although the length of the day would change, the length of the month would not! The Moon hasn't stopped rotating around the Earth. However, without the Earth's rotation feeding it tidal energy, the Moon *would* stop drifting away from the Earth (as it is doing currently) and would start to slowly drift back toward us.

2 Although without the Coriolis force, it's anyone's guess which way they would spin.

In fact, the Moon—our faithful companion—would act to undo the damage Andrew's scenario caused. Right now, the Earth spins faster than the Moon, and our tides slow down the Earth's rotation while pushing the Moon away from us.[3] If we stopped rotating, the Moon would stop drifting away from us. Instead of slowing us down, its tides would accelerate our spin. Quietly, gently, the Moon's gravity would tug on our planet . . .

. . . and Earth would start turning again.

Q. What would happen if you tried to hit a baseball pitched at 90 percent the speed of light?

—Ellen McManis

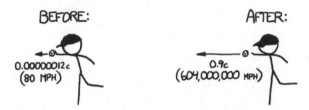

BEFORE:

0.000000012c
(80 MPH)

AFTER:

0.9c
(604,000,000 MPH)

Let's set aside the question of how we got the baseball moving that fast.
We'll suppose it's a normal pitch, except in the instant the pitcher releases the ball, it magically
accelerates to 0.9c. From that point onward, everything proceeds according to normal physics.

A. THE ANSWER TURNS OUT to be "a lot of things," and they all happen very quickly, and it doesn't end well for the batter (or the pitcher). I sat down with some physics books, a Nolan Ryan action figure, and a bunch of videotapes of nuclear tests and tried to sort it all out. What follows is my best guess at a nanosecond-by-nanosecond portrait.

The ball would be going so fast that everything else would be practically stationary. Even the molecules in the air would stand still. Air molecules would vibrate back and forth at a few hundred miles per hour, but the ball would be moving through them at 600 *million* miles per hour. This means that as far as the ball is concerned, they would just be hanging there, frozen.

The ideas of aerodynamics wouldn't apply here. Normally, air would flow around anything moving through it. But the air molecules in front of this ball wouldn't have time to be jostled out of the way. The ball would smack into them so hard that the atoms in the air molecules would actually fuse with the atoms in the ball's surface. Each collision would release a burst of gamma rays and scattered particles.[1]

These gamma rays and debris would expand outward in a bubble centered on the pitcher's mound. They would start to tear apart the molecules in the air, ripping the electrons from the nuclei and turning the air in the stadium into an expanding bubble of incandescent plasma. The wall of this bubble would approach the batter at about the speed of light—only slightly ahead of the ball itself.

The constant fusion at the front of the ball would push back on it, slowing it down, as if the ball were a rocket flying tail-first while firing its engines. Unfortunately, the ball would be going so fast that even the tremendous force from this ongoing thermonuclear explosion would barely slow it down at all. It would, however, start to eat away at the surface, blasting tiny fragments of the ball in all directions. These fragments would be going so fast that when they hit air molecules, they would trigger two or three more rounds of fusion.

After about 70 nanoseconds the ball would arrive at home plate. The batter wouldn't even have seen the pitcher let go of the ball, since the light carrying that information would arrive at about the same time the ball would. Collisions with the air would have eaten the ball away almost completely, and it would now be a bullet-shaped cloud of expanding plasma (mainly carbon, oxygen, hydrogen, and nitrogen) ramming into the air and triggering more fusion as it went. The shell of x-rays would hit the batter first, and a handful of nanoseconds later the debris cloud would hit.

---

1 After I initially published this article, MIT physicist Hans Rinderknecht contacted me to say that he'd simulated this scenario on their lab's computers. He found that early in the ball's flight, most of the air molecules were actually moving too quickly to cause fusion, and would pass right through the ball, heating it more slowly and uniformly than my original article described.

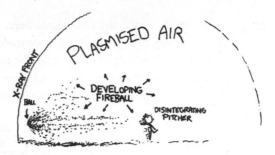

When it would reach home plate, the center of the cloud would still be moving at an appreciable fraction of the speed of light. It would hit the bat first, but then the batter, plate, and catcher would all be scooped up and carried backward through the backstop as they disintegrated. The shell of x-rays and superheated plasma would expand outward and upward, swallowing the backstop, both teams, the stands, and the surrounding neighborhood—all in the first microsecond.

Suppose you're watching from a hilltop outside the city. The first thing you would see would be a blinding light, far outshining the sun. This would gradually fade over the course of a few seconds, and a growing fireball would rise into a mushroom cloud. Then, with a great roar, the blast wave would arrive, tearing up trees and shredding houses.

Everything within roughly a mile of the park would be leveled, and a firestorm would engulf the surrounding city. The baseball diamond, now a sizable crater, would be centered a few hundred feet behind the former location of the backstop.

Major League Baseball Rule 6.08(b) suggests that in this situation, the batter would be considered "hit by pitch," and would be eligible to advance to first base.

Q. What if I took a swim in a typical spent nuclear fuel pool? Would I need to dive to actually experience a fatal amount of radiation? How long could I stay safely at the surface?

—Jonathan Bastien-Filiatrault

A. ASSUMING YOU'RE A REASONABLY good swimmer, you could probably survive treading water anywhere from 10 to 40 hours. At that point, you would black out from fatigue and drown. This is also true for a pool without nuclear fuel in the bottom.

Spent fuel from nuclear reactors is highly radioactive. Water is good for both radiation shielding and cooling, so fuel is stored at the bottom of pools for a couple of decades until it's inert enough to be moved into dry casks. We haven't really agreed on where to put those dry casks yet. One of these days we should probably figure that out.

Here's the geometry of a typical fuel storage pool:

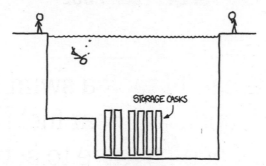

The heat wouldn't be a big problem. The water temperature in a fuel pool can in theory go as high as 50°C, but in practice it's generally between 25°C and 35°C—warmer than most pools but cooler than a hot tub.

The most highly radioactive fuel rods are those recently removed from a reactor. For the kinds of radiation coming off spent nuclear fuel, every 7 centimeters of water cuts the amount of radiation in half. Based on the activity levels provided by Ontario Hydro in this report, this would be the region of danger for fresh fuel rods:

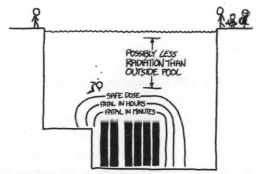

Swimming to the bottom, touching your elbows to a fresh fuel canister, and immediately swimming back up would probably be enough to kill you.

Yet outside the outer boundary, you could swim around as long as you wanted—the dose from the core would be less than the normal background dose you get walking around. In fact, as long as you were underwater, you would be shielded from most of that normal background dose. You may actually receive a lower dose of radiation treading water in a spent fuel pool than walking around on the street.

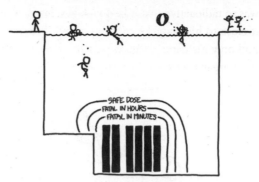

Remember: I am a cartoonist.
If you follow my advice on safety around nuclear materials,
you probably deserve whatever happens to you.

That's if everything goes as planned. If there's corrosion in the spent fuel rod casings, there may be some fission products in the water. They do a pretty good job of keeping the water clean, and it wouldn't hurt you to swim in it, but it's radioactive enough that it wouldn't be legal to sell it as bottled water.[1]

We know spent fuel pools can be safe to swim in because they're routinely serviced by human divers.

However, these divers have to be careful.

On August 31, 2010, a diver was servicing the spent fuel pool at the Leibstadt nuclear reactor in Switzerland. He spotted an unidentified length of tubing on the bottom of the pool and radioed his supervisor to ask what to do. He was told to put it in his tool basket, which he did. Due to bubble noise in the pool, he didn't hear his radiation alarm.

When the tool basket was lifted from the water, the room's radiation alarms went off. The basket was dropped back in the water and the diver left the pool. The diver's dosimeter badges showed that he'd received a higher-than-normal whole-body dose, and the dose in his right hand was extremely high.

The object turned out to be protective tubing from a radiation monitor in the reactor core, made highly radioactive by neutron flux. It had been accidentally

---

1 Which is too bad—it'd make a hell of an energy drink.

sheared off while a capsule was being closed in 2006. It sank to a remote corner of the pool, where it sat unnoticed for four years.

The tubing was so radioactive that if he'd tucked it into a tool belt or shoulder bag, where it sat close to his body, he could've been killed. As it was, the water protected him, and only his hand—a body part more resistant to radiation than the delicate internal organs—received a heavy dose.

So, as far as swimming safety goes, the bottom line is that you'd probably be OK, as long as you didn't dive to the bottom or pick up anything strange.

But just to be sure, I got in touch with a friend of mine who works at a research reactor, and asked him what he thought would happen to someone who tried to swim in their radiation containment pool.

"In *our* reactor?" He thought about it for a moment. "You'd die pretty quickly, before reaching the water, from gunshot wounds."

WEIRD (AND WORRYING) QUESTIONS
FROM THE WHAT IF? INBOX, #1

Q. Would it be possible to get your teeth to such a cold temperature that they would shatter upon drinking a hot cup of coffee?

—**Shelby Hebert**

Q. How many houses are burned down in the United States every year? What would be the easiest way to increase that number by a significant amount (say, at least 15%)?

—**Anonymous**

NEW YORK–STYLE TIME MACHINE

Q. I assume when you travel back in time you end up at the same spot on the Earth's surface. At least, that's how it worked in the *Back to the Future* movies. If so, what would it be like if you traveled back in time, starting in Times Square, New York, 1000 years? 10,000 years? 100,000 years? 1,000,000 years? 1,000,000,000 years? What about forward in time 1,000,000 years?

—**Mark Dettling**

1000 years back

Manhattan has been continuously inhabited for the past 3000 years, and was first settled by humans perhaps 9000 years ago.

In the 1600s, when Europeans arrived, the area was inhabited by the Lenape people.[1] The Lenape were a loose confederation of tribes who lived in what is now Connecticut, New York, New Jersey, and Delaware.

A thousand years ago, the area was probably inhabited by a similar collection of tribes, but those inhabitants lived half a millennium before European contact. They were as far removed from the Lenape of the 1600s as the Lenape of the 1600s are from the modern day.

To see what Times Square looked like before a city was there, we turn to a remarkable project called **Welikia,** which grew out of a smaller project called **Mannahatta.** The Welikia project has produced a detailed ecological map of the landscape in New York City at the time of the arrival of Europeans.

The interactive map, available online at *welikia.org,* is a fantastic snapshot of a different New York. In 1609, the island of Manhattan was part of a landscape of rolling hills, marshes, woodlands, lakes, and rivers.

The Times Square of 1000 years ago may have looked ecologically similar to the Times Square described by Welikia. Superficially, it probably resembled the old-growth forests that are still found in a few locations in the northeastern US. However, there would be some notable differences.

There would be more large animals 1000 years ago. Today's disconnected patchwork of northeastern old-growth forests is nearly free of large predators; we have some bears, few wolves and coyotes, and virtually no mountain lions. (Our deer populations, on the other hand, have exploded, thanks in part to the removal of large predators.)

The forests of New York 1000 years ago would be full of chestnut trees. Before a blight passed through in the early twentieth century, the hardwood forests of eastern North America were about 25 percent chestnut. Now, only their stumps survive.

You can still come across these stumps in New England forests today. They periodically sprout new shoots, only to see them wither as the blight takes hold. Someday, before too long, the last of the stumps will die.

---

1 Also known as the Delaware.

Wolves would be common in the forests, especially as you moved inland. You might also encounter mountain lions[2,3,4,5,6] and passenger pigeons.[7]

There's one thing you would *not* see: earthworms. There were no earthworms in New England when the European colonists arrived. To see the reason for the worms' absence, let's take our next step into the past.

10,000 years back

The Earth of 10,000 years ago was just emerging from a deep cold period.

The great ice sheets that covered New England had departed. As of 22,000 years ago, the southern edge of the ice was near Staten Island, but by 18,000 years ago it had retreated north past Yonkers.[8] By the time of our arrival, 10,000 years ago, the ice had largely withdrawn across the present-day Canadian border.

The ice sheets scoured the landscape down to bedrock. Over the next 10,000 years, life crept slowly back northward. Some species moved north faster than others; when Europeans arrived in New England, earthworms had not yet returned.

As the ice sheets withdrew, large chunks of ice broke off and were left behind.

---

2 Also known as cougars.
3 Also known as pumas.
4 Also known as catamounts.
5 Also known as panthers.
6 Also known as painted cats.
7 Although you might not see the clouds of trillions of pigeons encountered by European settlers. In his book *1491*, Charles C. Mann argues that the huge flocks seen by European settlers may have been a symptom of a chaotic ecosystem perturbed by the arrival of smallpox, bluegrass, and honeybees.
8 That is, the current site of Yonkers. It probably wasn't called "Yonkers" then, since "Yonkers" is a Dutch-derived name for a settlement dating to the late 1600s. However, some argue that a site called "Yonkers" has always existed, and in fact predates humans and the Earth itself. I mean, I guess it's just me who argues that, but I'm very vocal.

When these chunks melted, they left behind water-filled depressions in the ground called **kettlehole ponds.** Oakland Lake, near the north end of Springfield Boulevard in Queens, is one of these kettlehole ponds. The ice sheets also dropped boulders they'd picked up on their journey; some of these rocks, called **glacial erratics,** can be found in Central Park today.

Below the ice, rivers of meltwater flowed at high pressure, depositing sand and gravel as they went. These deposits, which remain as ridges called **eskers,** crisscross the landscape in the woods outside my home in Boston. They are responsible for a variety of odd landforms, including the world's only vertical U-shaped riverbeds.

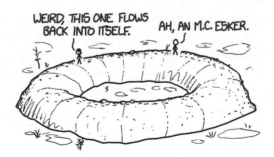

100,000 years back

The world of 100,000 years ago might have looked a lot like our own.[9] We live in an era of rapid, pulsating glaciations, but for 10,000 years our climate has been stable[10] and warm.

A hundred thousand years ago, Earth was near the end of a similar period of

---

9 Though with fewer billboards.
10 Well, *had* been. We're putting a stop to that.

climate stability. It was called the **Sangamon interglacial,** and it probably supported a developed ecology that would look familiar to us.

The coastal geography would be totally different; Staten Island, Long Island, Nantucket, and Martha's Vineyard were all berms pushed up by the most recent bulldozer-like advance of the ice. A hundred millennia ago, different islands dotted the coast.

Many of today's animals would be found in those woods—birds, squirrels, deer, wolves, black bears—but there would be a few dramatic additions. To learn about those, we turn to the mystery of the pronghorn.

The modern pronghorn (American antelope) presents a puzzle. It's a fast runner   in fact, it's much faster than it needs to be. It can run at 55 mph, and sustain that speed over long distances. Yet its fastest predators, wolves and coyotes, barely break 35 mph in a sprint. Why did the pronghorn evolve such speed?

The answer is that the world in which the pronghorn evolved was a much more dangerous place than ours. A hundred thousand years ago, North American woods were home to *Canis dirus* (the dire wolf), *Arctodus* (the short-faced bear), and *Smilodon fatalis* (sabre-toothed cat), each of which may have been faster and deadlier than modern predators. All died out in the Quaternary extinction event, which occured shortly after the first humans colonized the continent.[II]

If we go back a little further, we will meet another frightening predator.

1,000,000 years back

A million years ago, before the most recent great episode of glaciations, the world was fairly warm. It was the middle of the Quaternary period; the great modern ice ages had begun several million years earlier, but there had been a lull in the advance and retreat of the glaciers, and the climate was relatively stable.

The predators we met earlier, the fleet-footed creatures who may have preyed on the pronghorn, were joined by another terrifying carnivore, a long-limbed hyena that resembled a modern wolf. Hyenas were mainly found in Africa and Asia, but when the sea level fell, one species crossed the Bering Strait into North America. Because it was the only hyena to do so, it was given the name *Chasmaporthetes*, which means "the one who saw the canyon."

Next, Mark's question takes us on a great leap backward in time.

II If anyone asks, total coincidence.

1,000,000,000 years back

A billion years ago, the continental plates were pushed together into one great supercontinent. This was not the well-known supercontinent **Pangea**—it was Pangea's predecessor, **Rodinia.** The geologic record is spotty, but our best guess is that it looked something like this:

In the time of Rodinia, the bedrock that now lies under Manhattan had yet to form, but the deep rocks of North America were already old. The part of the continent that is now Manhattan was probably an inland region connected to what is now Angola and South Africa.

In this ancient world, there were no plants and no animals. The oceans were full of life, but it was simple single-cellular life. On the surface of the water were mats of blue-green algae.

These unassuming critters are the deadliest killers in the history of life.

Blue-green algae, or **cyanobacteria,** were the first photosynthesizers. They breathed in carbon dioxide and breathed out oxygen. Oxygen is a volatile gas; it causes iron to rust (oxidation) and wood to burn (vigorous oxidation). When cyanobacteria first appeared, the oxygen they breathed out was toxic to nearly all other forms of life. The resulting extinction is called the **oxygen catastrophe.**

After the cyanobacteria pumped Earth's atmosphere and water full of toxic oxygen, creatures evolved that took advantage of the gas's volatile nature to enable new biological processes. We are the descendants of those first oxygen-breathers.

Many details of this history remain uncertain; the world of a billion years ago is difficult to reconstruct. But Mark's question now takes us into an even more uncertain domain: the future.

1,000,000 years forward

Eventually, humans will die out. Nobody knows when,[12] but nothing lives forever. Maybe we'll spread to the stars and last for billions or trillions of years. Maybe civilization will collapse, we'll all succumb to disease and famine, and the last of us will be eaten by cats. Maybe we'll all be killed by nanobots hours after you read this sentence. There's no way to know.

A million years is a long time. It's several times longer than *Homo sapiens* has existed, and a hundred times longer than we've had written language. It seems reasonable to assume that however the human story plays out, in a million years it will have exited its current stage.

Without us, Earth's geology will grind on. Winds and rain and blowing sand will dissolve and bury the artifacts of our civilization. Human-caused climate change will probably delay the start of the next glaciation, but we haven't ended the cycle of ice ages. Eventually, the glaciers will advance again. A million years from now, few human artifacts will remain.

Our most lasting relic will probably be the layer of plastic we've deposited across the planet. By digging up oil, processing it into durable and long-lasting polymers, and spreading it across the Earth's surface, we've left a fingerprint that could outlast everything else we do.

Our plastic will become shredded and buried, and perhaps some microbes will learn to digest it, but in all likelihood, a million years from now, an out-of-place layer of processed hydrocarbons—transformed fragments of our shampoo bottles and shopping bags—will serve as a chemical monument to civilization.

The far future

The Sun is gradually brightening. For three billion years, a complex system of feedback loops has kept the Earth's temperature relatively stable as the Sun has grown steadily warmer.

In a billion years, these feedback loops will have given out. Our oceans, which nourished life and kept it cool, will have turned into its worst enemy. They will have boiled away in the hot Sun, surrounding the planet with a thick blanket of

---

12 If you do, email me.

water vapor and causing a runaway greenhouse effect. In a billion years, Earth will become a second Venus.

As the planet heats up, we may lose our water entirely and acquire a rock vapor atmosphere, as the crust itself begins to boil. Eventually, after several billion more years, we will be consumed by the expanding Sun.

The Earth will be incinerated, and many of the molecules that made up Times Square will be blasted outward by the dying Sun. These dust clouds will drift through space, perhaps collapsing to form new stars and planets.

If humans escape the solar system and outlive the Sun, our descendants may someday live on one of these planets. Atoms from Times Square, cycled through the heart of the Sun, will form our new bodies.

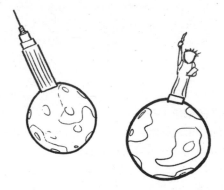

One day, either we will all be dead, or we will all be New Yorkers.

SOUL MATES

Q. What if everyone actually had only one soul mate, a random person somewhere in the world?

—Benjamin Staffin

- -

A. **WHAT A NIGHTMARE THAT** would be.

There are a lot of problems with the concept of a single random soul mate. As Tim Minchin put it in his song "If I Didn't Have You":

Your love is one in a million;
You couldn't buy it at any price.
But of the 9.999 hundred thousand other loves,
Statistically, some of them would be equally nice.

But what if we did have one randomly assigned perfect soul mate, and we *couldn't* be happy with anyone else? Would we find each other?

We'll assume your soul mate is chosen at birth. You don't know anything about who or where they are, but—as in the romantic cliché—you recognize each other the moment your eyes meet.

Right away, this would raise a few questions. For starters, would your soul mate even still be alive? A hundred billion or so humans have ever lived, but only seven billion are alive now (which gives the human condition a 93 percent mortality rate). If we were all paired up at random, 90 percent of our soul mates would be long dead.

That sounds horrible. But wait, it gets worse: A simple argument shows we can't limit ourselves just to past humans; we have to include an unknown number of future humans as well. See, if your soul mate is in the distant past, then it also has to be possible for soul mates to be in the distant future. After all, *your* soul mate's soul mate is.

So let's assume your soul mate lives at the same time as you. Furthermore, to keep things from getting creepy, we'll assume they're within a few years of your age. (This is stricter than the standard age-gap creepiness formula,[1] but if we assume a 30-year-old and a 40-year-old can be soul mates, then the creepiness rule is violated if they accidentally meet 15 years earlier.) With the same-age restriction, most of us would have a pool of around half a billion potential matches.

But what about gender and sexual orientation? And culture? And language? We could keep using demographics to try to narrow things down further, but we'd be drifting away from the idea of a random soul mate. In our scenario, you wouldn't know *anything* about who your soul mate was until you looked into their eyes. Everybody would have only one orientation: toward their soul mate.

The odds of running into your soul mate would be incredibly small. The number of strangers we make eye contact with each day can vary from almost none (shut-ins or people in small towns) to many thousands (a police officer in Times Square), but let's suppose you lock eyes with an average of a few dozen new strangers each day. (I'm pretty introverted, so for me that's definitely a generous estimate.) If 10 percent of them are close to your age, that would be around 50,000 people in a lifetime. Given that you have 500,000,000 potential soul mates, it means you would find true love only in one lifetime out of 10,000.

---

1 xkcd, "Dating pools," *http://xkcd.com/314.*

With the threat of dying alone looming so prominently, society could restructure to try to enable as much eye contact as possible. We could put together massive conveyer belts to move lines of people past each other . . .

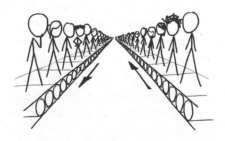

...but if the eye contact effect works over webcams, we could just use a modified version of ChatRoulette.

If everyone used the system for eight hours a day, seven days a week, and if it takes you a couple of seconds to decide if someone's your soul mate, this system could—in theory—match everyone up with their soul mates in a few decades. (I modeled a few simple systems to estimate how quickly people would pair off and drop out of the singles pool. If you want to try to work through the math for a particular setup, you might start by looking at derangement problems.)

In the real world, many people have trouble finding any time at all for romance—few could devote two decades to it. So maybe only rich kids would be able to afford to sit around on SoulMateRoulette. Unfortunately for the proverbial 1 percent, most of their soul mates would be found in the other 99 percent. If only 1 percent of the wealthy used the service, then 1 percent of that 1 percent would find their match through this system—one in 10,000.

The other 99 percent of the 1 percent[2] would have an incentive to get more people into the system. They might sponsor charitable projects to get computers to the rest of the world—a cross between One Laptop Per Child and OKCupid. Careers like "cashier" and "police officer in Times Square" would become high-status prizes because of the eye contact potential. People would flock to cities and public gathering places to find love—just as they do now.

But even if a bunch of us spent years on SoulMateRoulette, another bunch of us managed to hold jobs that offered constant eye contact with strangers, and the rest of us just hoped for luck, only a small minority of us would ever find true love. The rest of us would be out of luck.

Given all the stress and pressure, some people would fake it. They'd want to join the club, so they'd get together with another lonely person and stage a fake soul mate encounter. They'd marry, hide their relationship problems, and struggle to present a happy face to their friends and family.

A world of random soul mates would be a lonely one. Let's hope that's not what we live in.

---

2 "We are the zero point nine nine percent!"

LASER POINTER

Q. If every person on Earth aimed a laser pointer at the Moon at the same time, would it change color?
—Peter Lipowicz

A. NOT IF WE USED regular laser pointers.

The first thing to consider is that not everyone can see the Moon at once. We could gather everyone in one spot, but let's just pick a time when the Moon is visible to as many people as possible. Since about 75 percent of the world's population lives between 0°E and 120°E, we should try this while the Moon is somewhere over the Arabian Sea.

We could try to illuminate either a new moon or a full moon. The new moon is darker, making it easier to see our lasers. But the new moon is a trickier target, because it's mostly visible during the day—washing out the effect.

Let's pick a quarter moon, so we can compare the effect of our lasers on the dark and light sides.

Here's our target.

The typical red laser pointer is about 5 milliwatts, and a good one would have a tight enough beam to hit the Moon—though it'd be spread out over a large fraction of the surface when it got there. The atmosphere would distort the beam a bit, and absorb some of it, but most of the light would make it.

Let's assume everyone has steady enough aim to hit the Moon, but no more than that, and the light spreads evenly across the surface.

Half an hour after midnight (GMT), everyone aims and presses the button. This is what happened:

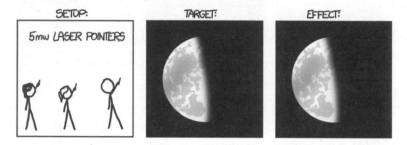

Well, that's disappointing.

It makes sense, though. Sunlight bathes the Moon in a bit over a kilowatt of energy per square meter. Since the Moon's cross-sectional area is around 10^{13} square meters, it's bathed in about 10^{16} watts of sunlight—10 petawatts, or 2 megawatts per person—far outshining our 5-milliwatt laser pointers. There are varying efficiencies in each part of this system, but none of it changes that basic equation.

A 1-watt laser is an extremely dangerous thing. It's not just powerful enough to blind you—it's capable of burning skin and setting things on fire. Obviously, they're not legal for consumer purchase in the US.

Just kidding! You can pick one up for $300. Just do a search for "1-watt hand-held laser."

So, suppose we spend the $2 trillion to buy 1-watt green lasers for everyone. (Memo to presidential candidates: This policy would win my vote.) In addition to being more powerful, green laser light is nearer to the middle of the visible spectrum, so the eye is more sensitive to it and it seems brighter.

Here's the effect:

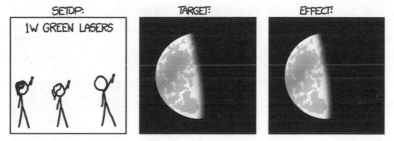

Dang.

The laser pointers we're using put out about 150 lumens of light (more than most flashlights) in a beam 5 arc-minutes wide. This lights up the surface of the Moon with about half a lux of illumination—compared to about 130,000 lux from the sun. (Even if we aimed them all perfectly, it would result in only half a dozen lux over about 10 percent of the Moon's face.)

By comparison, the full moon lights up the Earth's surface with about 1 lux of illumination—which means that not only would our lasers be too weak to see from Earth, but if you were standing on the Moon, the laser light on the landscape would be fainter than moonlight is to us on Earth.

With advances in lithium batteries and LED technology over the last ten years, the high-performance flashlight market has exploded. But it's clear that flashlights aren't gonna cut it. So let's skip past all of that and give everyone a Nightsun.

You may not recognize the name, but chances are you've seen one in operation: It's the searchlight mounted on police and Coast Guard helicopters. With an output on the order of 50,000 lumens, it's capable of turning a patch of ground from night to day.

The beam is several degrees wide, so we would want some focusing lenses to get it down to the half-degree needed to hit the Moon.

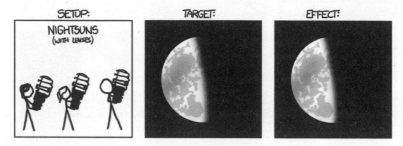

It's hard to see, but we're making progress! The beam is providing 20 lux of illumination, outshining the ambient light on the night half by a factor of two! However, it's quite hard to see, and it certainly hasn't affected the light half.

WHAT IF WE TRIED
MORE POWER?

Let's swap out each Nightsun for an IMAX projector array—a 30,000-watt pair of water-cooled lamps with a combined output of over a million lumens.

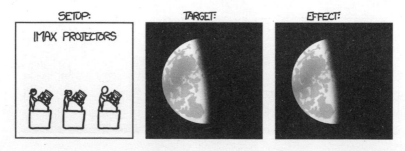

Still barely visible.

At the top of the Luxor Hotel in Las Vegas is the most powerful spotlight on Earth. Let's give one of them to everyone.

Oh, and let's add a lens array to each so the entire beam is focused on the Moon:

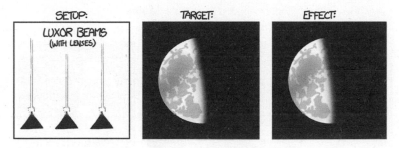

Our light is definitely visible, so we've accomplished our goal! Good job, team.

Well . . .

The Department of Defense has developed megawatt lasers, designed for destroying incoming missiles in mid-flight.

The Boeing YAL-1 was a megawatt-class chemical oxygen iodine laser mounted in a 747. It was an infrared laser, so it wasn't directly visible, but we can imagine building a visible-light laser with similar power.

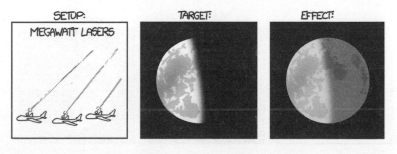

Finally, we've managed to match the brightness of sunlight!

We're also drawing 5 petawatts of power, which is double the world's average electricity consumption.

Okay, let's mount a megawatt laser on every square meter of Asia's surface. Powering this array of 50 trillion lasers would use up Earth's oil reserves in approximately two minutes, but for those two minutes, the Moon would look like this:

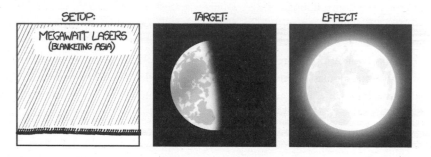

The Moon would shine as brightly as the midmorning sun, and by the end of the two minutes, the lunar regolith would be heated to a glow.

Okay, let's step even more firmly outside the realm of plausibility.

The most powerful laser on Earth is the confinement beam at the National Ignition Facility, a fusion research laboratory. It's an ultraviolet laser with an output of 500 terawatts. However, it fires only in single pulses lasting a few nanoseconds, so the total energy delivered is equivalent to about a quarter-cup of gasoline.

Let's imagine we somehow found a way to power and fire it continuously, gave one to everyone, and pointed them all at the Moon. Unfortunately, the laser energy flow would turn the atmosphere to plasma, instantly igniting the Earth's surface and killing us all. But let's assume that the lasers somehow pass through the atmosphere without interacting.

Under those circumstances, it turns out Earth would *still* catch fire. The reflected light from the Moon would be four thousand times brighter than the noonday sun. Moonlight would become bright enough to boil away Earth's oceans in less than a year.

But forget the Earth—what would happen to the Moon?

The laser itself would exert enough radiation pressure to accelerate the Moon at about one ten millionth of a gee. This acceleration wouldn't be noticeable in the short term, but over the years, it would add up to enough to push it free from Earth orbit . . .

. . . if radiation pressure were the only force involved.

Forty megajoules of energy is enough to vaporize a kilogram of rock. Assuming Moon rocks have an average density of about 3 kg/liter, the lasers would pump out enough energy to vaporize 4 meters of lunar bedrock per second:

$$\frac{5 \text{ billion people} \times 500 \frac{\text{terawatts}}{\text{person}}}{\pi \times \text{Moon radius}^2} \times 20 \, \frac{\text{megajoules}}{\text{kilogram}} \times 3 \, \frac{\text{kilograms}}{\text{liter}} \approx 4 \, \frac{\text{meters}}{\text{second}}$$

However, the actual lunar rock wouldn't evaporate that fast—for a reason that turns out to be very important.

When a chunk of rock is vaporized, it doesn't just disappear. The surface layer of the Moon becomes a plasma, but that plasma would still block the path of the beam.

Our laser would keep pouring more and more energy into the plasma, and the plasma would keep getting hotter and hotter. The particles would bounce off each other, slam into the surface of the Moon, and eventually blast into space at a terrific speed.

This flow of material effectively turns the entire surface of the Moon into a rocket engine—and a surprisingly efficient one, too. Using lasers to blast off surface material like this is called laser ablation, and it turns out to be a promising method for spacecraft propulsion.

The Moon is massive, but slowly and surely the rock plasma jet would begin to push it away from the Earth. (The jet would also scour the face of the Earth clean and destroy the lasers, but we're pretending that they're invulnerable.) The plasma would also physically tear away the lunar surface, a complicated interaction that's tricky to model.

But if we make the wild guess that the particles in the plasma exit at an average speed of 500 kilometers per second, then it will take a few months for the Moon to be pushed out of range of our laser. It would keep most of its mass, but escape Earth's gravity and enter a lopsided orbit around the sun.

Technically, the Moon wouldn't become a new planet, under the IAU definition of a planet. Since its new orbit would cross Earth's, it would be considered a dwarf planet like Pluto. This Earth-crossing orbit would lead to periodic unpredictable orbital perturbation. Eventually it would either be slingshotted into the Sun, ejected toward the outer solar system, or slammed into one of the planets—quite possibly ours. I think we can all agree that in this case, we'd deserve it.

Scorecard:

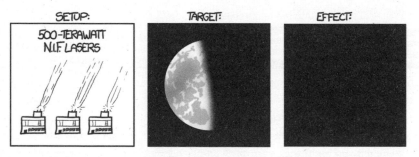

And that, at last, would be enough power.

PERIODIC WALL OF THE ELEMENTS

Q. What would happen if you made a periodic table out of cube-shaped bricks, where each brick was made of the corresponding element?

—Andy Connolly

- -

A. THERE ARE PEOPLE WHO collect elements. These collectors try to gather physical samples of as many of the elements as possible into periodic-table-shaped display cases.[1]

Of the 118 elements, 30 of them—like helium, carbon, aluminum, and iron—can be bought in pure form in local retail stores. Another few dozen can be scavenged by taking things apart (you can find tiny americium samples in smoke detectors). Others can be ordered over the Internet.

All in all, it's possible to get samples of about 80 of the elements—90, if you're willing to take some risks with your health, safety, and arrest record. The rest are too radioactive or short-lived to collect more than a few atoms of them at once.

But what if you *did*?

The periodic table of the elements has seven rows.[2]

---

1 Think of the elements as dangerous, radioactive, short-lived Pokémon.
2 An eighth row may be added by the time you read this. And if you're reading this in the year 2038, the periodic table has ten rows but all mention or discussion of it is banned by the robot overlords.

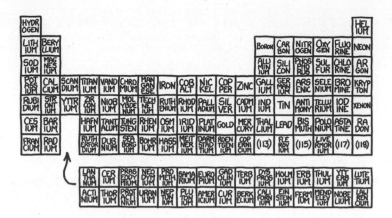

- You could stack the top two rows without much trouble.
- The third row would burn you with fire.
- The fourth row would kill you with toxic smoke.
- The fifth row would do all that stuff PLUS give you a mild dose of radiation.
- The sixth row would explode violently, destroying the building in a cloud of radioactive, poisonous fire and dust.
- Do not build the seventh row.

We'll start from the top. The first row is simple, if boring:

The cube of hydrogen would rise upward and disperse, like a balloon without a balloon. The same goes for helium.

The second row is trickier.

The lithium would immediately tarnish. The beryllium is pretty toxic, so you should handle it carefully and avoid getting any dust in the air.

The oxygen and nitrogen drift around, slowly dispersing. The neon floats away.[3]

The pale yellow fluorine gas would spread across the ground. Fluorine is the most reactive, corrosive element in the periodic table. Almost any substance exposed to pure fluorine will spontaneously catch fire.

I spoke to organic chemist Derek Lowe about this scenario.[4] He said that the fluorine wouldn't react with the neon, and "would observe a sort of armed truce with the chlorine, but everything else, sheesh." Even with the later rows, the fluorine would cause problems as it spread, and if it came in contact with any moisture, it would form corrosive hydrofluoric acid.

If you breathed even a trace amount, it would seriously damage or destroy your nose, lungs, mouth, eyes, and eventually the rest of you. You would definitely need a gas mask. Keep in mind that fluorine eats through a lot of potential mask materials, so you would want to test it first. Have fun!

On to the third row!

Half of the data here is from the CRC Handbook of Chemistry and Physics *and the other half is from* Look Around You.

---

3 That is, assuming that they're in diatomic form (e.g. O_2 and N_2). If the cube is in the form of single atoms, they'll instantly combine, heating to thousands of degrees as they do.
4 Lowe is the author of the great drug research blog *In the Pipeline.*

The big troublemaker here is phosphorus. Pure phosphorus comes in several forms. Red phosphorus is reasonably safe to handle. White phosphorus spontaneously ignites on contact with air. It burns with hot, hard-to-extinguish flames and is, in addition, quite poisonous.[5]

The sulfur wouldn't be a problem under normal circumstances; at worst, it would smell bad. However, our sulfur is sandwiched between burning phosphorus on the left . . . and the fluorine and chlorine on the right. When exposed to pure fluorine gas, sulfur—like many substances—catches fire.

The inert argon is heavier than air, so it would just spread out and cover the ground. Don't worry about the argon. You have bigger problems.

The fire would produce all kinds of terrifying chemicals with names like sulfur hexafluoride. If you're doing this inside, you'd be choked by toxic smoke and your building might burn down.

And that's only row three. On to row four!

"Arsenic" sounds scary. The reason it sounds scary is a good one: It's toxic to virtually all forms of complex life.

Sometimes this kind of panic over scary chemicals is disproportionate; there are trace amounts of natural arsenic in all our food and water, and we handle those fine. This is not one of those times.

The burning phosphorus (now joined by burning potassium, which is similarly prone to spontaneous combustion) could ignite the arsenic, releasing large amounts of arsenic trioxide. That stuff is pretty toxic. Don't inhale.

This row would also produce hideous odors. The selenium and bromine

---

5 A property that has led to its controversial use in incendiary artillery shells.

would react vigorously, and Lowe says that burning selenium "can make sulfur smell like Chanel."

If the aluminum survived the fire, a strange thing would happen to it. The melting gallium under it would soak into the aluminum, disrupting its structure and causing it to become as soft and weak as wet paper.[6]

The burning sulfur would spill into the bromine. Bromine is liquid at room temperature, a property it shares with only one other element—mercury. It's also pretty nasty stuff. The range of toxic compounds that would be produced by this blaze is, at this point, incalculably large. However, if you did this experiment from a safe distance, you might survive.

The fifth row contains something interesting: technetium 99, our first radioactive brick.

Technetium is the lowest-numbered element that has no stable isotopes. The dose from a 1-liter cube of the metal wouldn't be enough to be lethal in our experiment, but it's still substantial. If you spent all day wearing it as a hat—or breathed it in as dust—it could definitely kill you.

NOT A HAT

Techneteium aside, the fifth row would be a lot like the fourth.

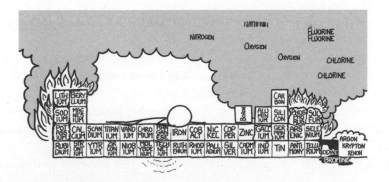

---

6 Search YouTube for "gallium infiltration" to see how strange this is.

On to the sixth row! No matter how careful you are, the sixth row would definitely kill you.

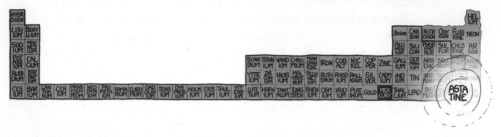

This version of the periodic table is a little wider than you might be used to, since we're inserting the lanthanide and actinide elements into rows 6 and 7. (These elements are normally shown separately from the main table to avoid making it too wide.)

The sixth row of the periodic table contains several radioactive elements, including promethium, polonium,[7] astatine, and radon. Astatine is the bad one.[8]

We don't know what astatine looks like, because, as Lowe put it, "that stuff just doesn't want to exist." It's so radioactive (with a half-life measured in hours) that any large piece of it would be quickly vaporized by its own heat. Chemists suspect that it has a black surface, but no one really knows.

There's no material safety data sheet for astatine. If there were, it would just be the word "NO" scrawled over and over in charred blood.

Our cube would, briefly, contain more astatine than has ever been synthesized. I say "briefly" because it would immediately turn into a column of superheated gas. The heat alone would give third-degree burns to anyone nearby, and the building would be demolished. The cloud of hot gas would rise rapidly into the sky, pouring out heat and radiation.

The explosion would be just the right size to maximize the amount of paperwork your lab would face. If the explosion were smaller, you could potentially cover it up. If it were larger, there would be no one left in the city to submit paperwork to.

Dust and debris coated in astatine, polonium, and other radioactive products

---

7 In 2006, an umbrella tipped with polonium-210 was used to murder former KGB officer Alexander Litvinenko.
8 Radon is the cute one.

would rain from the cloud, rendering the downwind neighborhood completely uninhabitable.

The radiation levels would be incredibly high. Given that it takes a few hundred milliseconds to blink, you would literally get a lethal dose of radiation in the blink of an eye.

You would die from what we might call "extremely acute radiation poisoning"—that is, you would be cooked.

The seventh row would be much worse.

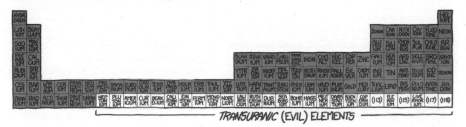

TRANSURANIC (EVIL) ELEMENTS

There are a whole bunch of weird elements along the bottom of the periodic table called **transuranic elements.** For a long time, many of them had place-holder names like "unununium," but gradually they're being assigned permanent names.

There's no rush, though, because most of these elements are so unstable that they can be created only in particle accelerators and don't exist for more than a few minutes. If you had 100,000 atoms of Livermorium (element 116), after a second you'd have one left—and a few hundred milliseconds later, that one would be gone, too.

Unfortunately for our project, the transuranic elements don't vanish quietly. They decay radioactively. And most of them decay into things that *also* decay. A cube of any of the highest-numbered elements would decay within seconds, releasing a tremendous amount of energy.

The result wouldn't be like a nuclear explosion—it *would* be a nuclear explosion. However, unlike a fission bomb, it wouldn't be a chain reaction—just a reaction. It would all happen at once.

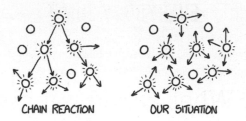

CHAIN REACTION   OUR SITUATION

The flood of energy would instantly turn you—and the rest of the periodic table—to plasma. The blast would be similar to that of a medium-sized nuclear detonation, but the radioactive fallout would be much, much worse—a veritable salad of everything on the periodic table turning into everything else as fast as possible.

A mushroom cloud would rise over the city. The top of the plume would reach up through the stratosphere, buoyed by its own heat. If you were in a populated area, the immediate casualties from the blast would be staggering, but the long-term contamination from the fallout would be even worse.

The fallout wouldn't be normal, everyday radioactive fallout[9]—it would be like a nuclear bomb that *kept exploding*. The debris would spread around the world, releasing thousands of times more radioactivity than the Chernobyl disaster. Entire regions would be devastated; the cleanup would stretch on for centuries.

While collecting things is certainly fun, when it comes to chemical elements, you do *not* want to collect them all.

MAYBE I COULD JUST DESTROY THE WORLD A *LITTLE*...

---

9 You know, the stuff we all shrug off.

EVERYBODY JUMP

Q. What would happen if everyone on Earth stood as close to each other as they could and jumped, everyone landing on the ground at the same instant?

—Thomas Bennett (and many others)

- -

A. THIS IS ONE OF the most popular questions submitted through my website. It's been examined before, including by *ScienceBlogs* and *The Straight Dope*. They cover the kinematics pretty well. However, they don't tell the whole story.

Let's take a closer look.

At the start of the scenario, the entire Earth's population has been magically transported together into one place.

This crowd takes up an area the size of Rhode Island. But there's no reason to use the vague phrase "an area the size of Rhode Island." This is our scenario; we can be specific. They're *actually* in Rhode Island.

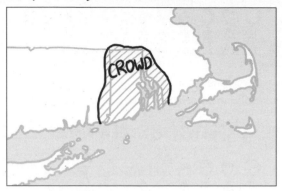

At the stroke of noon, everyone jumps.

As discussed elsewhere, it doesn't really affect the planet. Earth outweighs us by a factor of over ten trillion. On average, we humans can vertically jump maybe half a meter on a good day. Even if the Earth were rigid and responded instantly, it would be pushed down by less than an atom's width.

Next, everyone falls back to the ground.

Technically, this delivers a lot of energy into the Earth, but it's spread out over a large enough area that it doesn't do much more than leave footprints in a lot of gardens. A slight pulse of pressure spreads through the North American continental crust and dissipates with little effect. The sound of all those feet hitting the ground creates a loud, drawn-out roar lasting many seconds.

Eventually, the air grows quiet.

Seconds pass. Everyone looks around.

There are a lot of uncomfortable glances. Someone coughs.

A cell phone comes out of a pocket. Within seconds, the rest of the world's five billion phones follow. All of them—even those compatible with the region's towers—are displaying some version of "NO SIGNAL." The cell networks have all collapsed under the unprecedented load. Outside Rhode Island, abandoned machinery begins grinding to a halt.

The T. F. Green Airport in Warwick, Rhode Island, handles a few thousand passengers a day. Assuming they got things organized (including sending out scouting missions to retrieve fuel), they could run at 500 percent capacity for years without making a dent in the crowd.

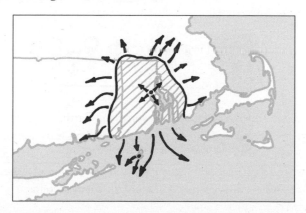

The addition of all the nearby airports doesn't change the equation much. Nor does the region's light rail system. Crowds climb on board container ships in the deep-water port of Providence, but stocking sufficient food and water for a long sea voyage proves a challenge.

Rhode Island's half-million cars are commandeered. Moments later, I-95, I-195, and I-295 become the sites of the largest traffic jam in the history of the planet. Most of the cars are engulfed by the crowds, but a lucky few get out and begin wandering the abandoned road network.

Some make it past New York or Boston before running out of fuel. Since the electricity is probably not on at this point, rather than find a working gas pump, it's easier to just abandon the car and steal a new one. Who can stop you? All the cops are in Rhode Island.

The edge of the crowd spreads outward into southern Massachusetts and Connecticut. Any two people who meet are unlikely to have a language in common, and almost nobody knows the area. The state becomes a chaotic patchwork of coalescing and collapsing social hierarchies. Violence is common. Everybody is hungry and thirsty. Grocery stores are emptied. Fresh water is hard to come by and there's no efficient system for distributing it.

Within weeks, Rhode Island is a graveyard of billions.

The survivors spread out across the face of the world and struggle to build a new civilization atop the pristine ruins of the old. Our species staggers on, but our population has been greatly reduced. Earth's orbit is completely unaffected—it spins along exactly as it did before our species-wide jump.

But at least now we know.

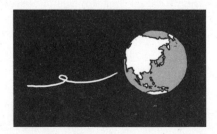

A MOLE OF MOLES

Q. What would happen if you were to gather a mole (unit of measurement) of moles (the small furry critter) in one place?

—Sean Rice

A. **THINGS GET A BIT** gruesome.

First, some definitions.

A mole is a unit. It's not a typical unit, though. It's really just a number—like "dozen" or "billion." If you have a mole of something, it means you have 602,214,129,000,000,000,000,000 of them (usually written 6.022×10^{23}). It's such a big number[1] because it's used for counting numbers of molecules, which there are a lot of.

THERE ARE TOO MANY MOLECULES.

---

[1] "One mole" is close to the number of atoms in a gram of hydrogen. It's also, by chance, a decent ballpark guess for the number of grains of sand on Earth.

A mole is also a type of burrowing mammal. There are a handful of types of moles, and some of them are truly horrifying.[2]

So what would a mole of moles—602,214,129,000,000,000,000,000 animals—look like?

First, let's start with wild approximations. This is an example of what might go through my head before I even pick up a calculator, when I'm just trying to get a sense of the quantities—the kind of calculation where 10, 1, and 0.1 are all close enough that we can consider them equal:

A mole (the animal) is small enough for me to pick up and throw.[citation needed] Anything I can throw weighs 1 pound. One pound is 1 kilogram. The number 602,214,129,000,000,000,000,000 looks about twice as long as a trillion, which means it's about a trillion trillion. I happen to remember that a trillion trillion kilograms is how much a planet weighs.

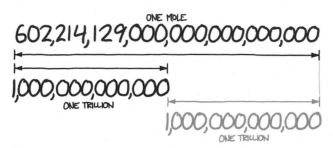

. . . if anyone asks, I did **not** tell you it was okay to do math like this.

That's enough to tell us that we're talking about a pile of moles on the scale of

2 http://en.wikipedia.org/wiki/File:Condylura.jpg

a planet. It's a pretty rough estimate, since it could be off by a factor of thousands in either direction.

Let's get some better numbers.

An eastern mole (*Scalopus aquaticus*) weighs about 75 grams, which means a mole of moles weighs:

$$(6.022 \times 10^{23}) \times 75g \approx 4.52 \times 10^{22} \, kg$$

That's a little over half the mass of our moon.

Mammals are largely water. A kilogram of water takes up a liter of volume, so if the moles weigh 4.52×10^{22} kilograms, they take up about 4.52×10^{22} liters of volume. You might notice that we're ignoring the pockets of space between the moles. In a moment, you'll see why.

The cube root of 4.52×10^{22} liters is 3562 kilometers, which means we're talking about a sphere with a radius of 2210 kilometers, or a cube 2213 miles on each edge.[3]

If these moles were released onto the Earth's surface, they'd fill it up to 80 kilometers deep—just about to the (former) edge of space:

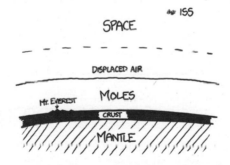

This smothering ocean of high-pressure meat would wipe out most life on the planet, which could—to reddit's horror—threaten the integrity of the DNS system. So doing this on Earth is definitely not an option.

Instead, let's gather the moles in interplanetary space. Gravitational attraction would pull them into a sphere. Meat doesn't compress very well, so it would

---

3 That's a neat coincidence I've never noticed before—a cubic mile happens to be almost exactly $4/3\pi$ cubic kilometers, so a sphere with a radius of X kilometers has the same volume as a cube that's X miles on each side.

undergo only a little bit of gravitational contraction, and we'd end up with a mole planet slightly larger than the Moon.

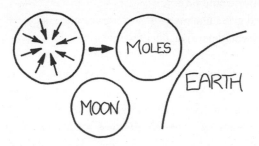

The moles would have a surface gravity of about one-sixteenth of Earth's—similar to that of Pluto. The planet would start off uniformly lukewarm—probably a bit over room temperature—and the gravitational contraction would heat the deep interior by a handful of degrees.

But this is where it gets weird.

The mole planet would be a giant sphere of meat. It would have a lot of latent energy (there are enough calories in the mole planet to support the Earth's current population for 30 billion years). Normally, when organic matter decomposes, it releases much of that energy as heat. But throughout the majority of the planet's interior, the pressure would be over 100 megapascals, which is high enough to kill all bacteria and sterilize the mole remains—leaving no microorganisms to break down the mole tissue.

Closer to the surface, where the pressure would be lower, there would be another obstacle to decomposition—the interior of a mole planet would be low in oxygen. Without oxygen, the usual decomposition couldn't happen, and the only bacteria that would be able to break down the moles would be those that don't require oxygen. While inefficient, this anaerobic decomposition can unlock quite a bit of heat. If continued unchecked, it would heat the planet to a boil.

But the decomposition would be self-limiting. Few bacteria can survive at temperatures above about 60°C, so as the temperature went up, the bacteria would die off, and the decomposition would slow. Throughout the planet, the mole bodies would gradually break down into kerogen, a mush of organic matter that would—if the planet were hotter—eventually form oil.

The outer surface of the planet would radiate heat into space and freeze.

Because the moles form a literal fur coat, when frozen they would insulate the interior of the planet and slow the loss of heat to space. However, the flow of heat in the liquid interior would be dominated by convection. Plumes of hot meat and bubbles of trapped gases like methane—along with the air from the lungs of the deceased moles—would periodically rise through the mole crust and erupt volcanically from the surface, a geyser of death blasting mole bodies free of the planet.

Eventually, after centuries or millennia of turmoil, the planet would calm and cool enough that it would begin to freeze all the way through. The deep interior would be under such high pressure that as it cooled, the water would crystallize out into exotic forms of ice such as ice III and ice V, and eventually ice II and ice IX.[4]

All told, this is a pretty bleak picture. Fortunately, there's a better approach.

I don't have any reliable numbers for global mole population (or small mammal biomass in general), but we'll take a shot in the dark and estimate that there are at least a few dozen mice, rats, voles, and other small mammals for every human.

There might be a billion habitable planets in our galaxy. If we colonized them, we'd certainly bring mice and rats with us. If just one in a hundred were populated with small mammals in numbers similar to Earth's, after a few million years—not long, in evolutionary time—the total number that have ever lived would surpass Avogadro's number.

If you want a mole of moles, build a spaceship.

---

4 No relation.

HAIR DRYER

Q. What would happen if a hair dryer with continuous power were turned on and put in an airtight $1 \times 1 \times 1$-meter box?

—Dry Paratroopa

- -

A. A TYPICAL HAIR DRYER draws 1875 watts of power.

All 1875 watts have to go somewhere. No matter what happens inside the box, if it's using 1875 watts of power, eventually there will be 1875 watts of heat flowing out.

This is true of any device that uses power, which is a handy thing to know. For example, people worry about leaving disconnected chargers plugged into the wall for fear that they're draining power. Are they right? Heat flow analysis provides a simple rule of thumb: If an unused charger isn't warm to the touch, it's using less than a penny of electricity a day. For a small smartphone charger, if it's not warm to the touch, it's using less than a penny a *year*. This is true of almost any powered device.[1]

But back to the box.

Heat will flow from the hair dryer out into the box. If we assume the dryer is indestructible, the interior of the box will keep getting hotter until the outer

---

1 Though not necessarily those plugged into a second device. If a charger is connected to something, like a smartphone or laptop, power can be flowing from the wall through the charger into the device.

surface reaches about 60°C (140°F). At that temperature, the box will be losing heat to the outside as fast as the hair dryer is adding it inside, and the system will be in equilibrium.

It's warmer than my parents! It's my new parents.

The equilibrium temperature will be a bit cooler if there's a breeze, or if the box is sitting on a wet or metallic surface that conducts away heat quickly.

If the box is made of metal, it will be hot enough to burn your hand if you touch it for more than five seconds. If it's wood, you can probably touch it for a while, but there's a danger that parts of the box in contact with the mouth of the hair dryer will catch fire.

The inside of the box will be like an oven. The temperature it reaches will depend on the thickness of the box wall; the thicker and more insulating the wall, the higher the temperature. It wouldn't take a very thick box to create temperatures high enough to burn out the hair dryer.

But let's assume it's an indestructible hair dryer. And if we have something as cool as an indestructible hair dryer, it seems like a shame to limit it to 1875 watts.

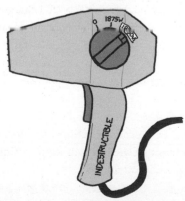

With 18,750 watts flowing out of the hair dryer, the surface of the box reaches over 200°C (475°F), as hot as a skillet on low-medium.

I wonder how high this dial goes.

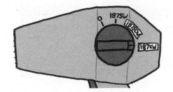

There's a distressing amount of space left on the dial.

The surface of the box is now 600°C, hot enough to glow a dim red.

If it's made of aluminium, the inside is starting to melt. If it's made of lead, the outside is starting to melt. If it's on a wood floor, the house is on fire. But it doesn't matter what's happening around it; the hair dryer is indestructible.

Two megawatts pumped into a laser is enough to destroy missiles.

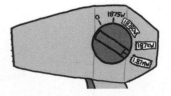

At 1300°C, the box is now about the temperature of lava.

One more notch.

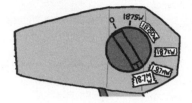

This hair dryer is probably not up to code.

Now 18 megawatts are flowing into the box.

The surface of the box reaches 2400°C. If it were steel, it would have melted by now. If it's made of something like tungsten, it might conceivably last a little longer.

Just one more, then we'll stop.

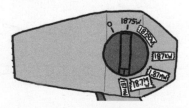

This much power—187 megawatts—is enough to make the box glow white. Not a lot of materials can survive these conditions, so we'll have to assume the box is indestructible.

The floor is made of lava.

Unfortunately, the floor isn't.

Before it can burn its way through the floor, someone throws a water balloon under it. The burst of steam launches the box out the front door and onto the sidewalk.[2]

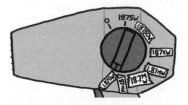

We're at 1.875 gigawatts (I lied about stopping). According to *Back to the Future,* the hair dryer is now drawing enough power to travel back in time.

---

2 Note: If you're ever trapped with me in a burning building, and I suggest an idea for how we could escape the situation, it's probably best to ignore me.

The box is blindingly bright, and you can't get closer than a few hundred meters due to the intense heat. It sits in the middle of a growing pool of lava. Anything within 50–100 meters bursts into flame. A column of heat and smoke rise high into the air. Periodic explosions of gas beneath the box launch it into the air, and it starts fires and forms a new lava pool where it lands.

We keep turning the dial.

At 18.7 gigawatts, the conditions around the box are similar to those on the pad during a space shuttle launch. The box begins to be tossed around by the powerful updrafts it's creating.

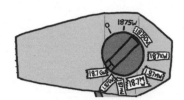

In 1914, H. G. Wells imagined devices like this in his book *The World Set Free*. He wrote of a type of bomb that, instead of exploding once, exploded *continuously,* a slow-burn inferno that started inextinguishable fires in the hearts of cities. The story eerily foreshadowed the development, 30 years later, of nuclear weapons.

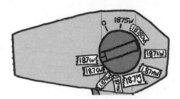

The box is now soaring through the air. Each time it nears the ground, it superheats the surface, and the plume of expanding air hurls it back into the sky.

The outpouring of 1.875 terawatts is like a house-sized stack of TNT going off *every second.*

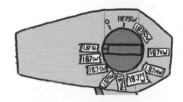

A trail of firestorms—massive conflagrations that sustain themselves by creating their own wind systems—winds its way across the landscape.

A new milestone: The hair dryer is now, impossibly, consuming more power than every other electrical device on the planet combined.

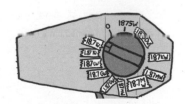

The box, soaring high above the surface, is putting out energy equivalent to three Trinity tests every *second.*

At this point, the pattern is obvious. This thing is going to skip around the atmosphere until it destroys the planet.

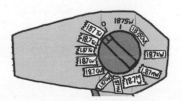

Let's try something different.

We turn the dial to zero as the box is passing over northern Canada. Rapidly cooling, it plummets to Earth, landing in Great Bear Lake with a plume of steam.

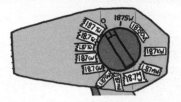

And then . . .

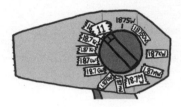

In this case, that's 11 petawatts.

A brief story:

The official record for the fastest manmade object is the Helios 2 probe, which reached about 70 km/s in a close swing around the Sun. But it's possible the actual holder of that title is a two-ton metal manhole cover.

The cover sat atop a shaft at an underground nuclear test site operated by Los Alamos as part of Operation Plumbbob. When the 1-kiloton nuke went off below, the facility effectively became a nuclear potato cannon, giving the cap a gigantic kick. A high-speed camera trained on the lid caught only one frame of it moving upward before it vanished—which means it was moving at a minimum of 66 km/s. The cap was never found.

Now, 66 km/s is about six times escape velocity, but contrary to common speculation, it's unlikely the cap ever reached space. Newton's impact depth approximation suggests that it was either destroyed completely by impact with the air or slowed and fell back to Earth.

When we turn it back on, our reactivated hair dryer box, bobbing in lake water, undergoes a similar process. The heated steam below it expands outward, and as the box rises into the air, the entire surface of the lake turns to steam. The steam, heated to a plasma by the flood of radiation, accelerates the box faster and faster.

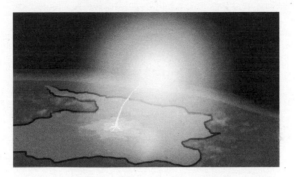

Photo courtesy of Commander Hadfield

Rather than slam into the atmosphere like the manhole cover, the box flies through a bubble of expanding plasma that offers little resistance. It exits the atmosphere and continues away, slowly fading from second sun to dim star. Much of the Northwest Territories is burning, but the Earth has survived.

However, a few may wish we hadn't.

WEIRD (AND WORRYING) QUESTIONS FROM THE WHAT IF? INBOX, #2

Q. Would dumping anti-matter into the Chernobyl reactor when it was melting down stop the meltdown?

—AJ

A.J., IN RECOGNITION OF YOUR CHERNOBYL RESPONSE EFFORTS, WE AWARD YOU THE "FOR GOD'S SAKE, WHAT WERE YOU *THINKING?!*" AWARD.

IT'S SHAPED LIKE A VHS TAPE OF THE *STAR WARS HOLIDAY SPECIAL*.

Q. Is it possible to cry so much you dehydrate yourself?

—Karl Wildermuth

...KARL, IS EVERYTHING OK?

THE LAST HUMAN LIGHT

Q. If every human somehow simply disappeared from the face of the Earth, how long would it be before the last artificial light source would go out?

—**Alan**

A. THERE WOULD BE A lot of contenders for the "last light" title.

The superb 2007 book *The World Without Us*, by Alan Weisman, explored in great detail what would happen to Earth's houses, roads, skyscrapers, farms, and animals if humans suddenly vanished. A 2008 TV series called *Life After People* investigated the same premise. However, neither of them answered this particular question.

We'll start with the obvious: Most lights wouldn't last long, because the major power grids would go down relatively fast. Fossil fuel plants, which supply the vast majority of the world's electricity, require a steady supply of fuel, and their supply chains do involve humans making decisions.

ON AUGUST 4TH, 2017, SKYNET WAS BROUGHT ONLINE AND PUT IN CHARGE OF OUR POWER PLANT FUEL PURCHASING DECISIONS.

ON AUGUST 29TH, IT BECAME SELF-AWARE AND DECIDED TO DESTROY HUMANITY.

FORTUNATELY, ALL IT COULD DO WAS REFUSE TO BUY FUEL.

EVENTUALLY, SOMEONE TURNED IT BACK OFF.

OH WELL.

Without people, there would be less demand for power, but our thermostats would still be running. As coal and oil plants started shutting down in the first few hours, other plants would need to take up the slack. This kind of situation is difficult to handle even *with* human guidance. The result would be a rapid series of cascade failures, leading to a blackout of all the major power grids.

However, plenty of electricity comes from sources not tied to the major power grids. Let's take a look at a few of those, and when each one might turn off.

Diesel generators

Many remote communities, like those on far-flung islands, get their power from diesel generators. These can continue to operate until they run out of fuel, which in most cases could be anywhere from days to months.

Geothermal plants

Generating stations that don't need a human-provided fuel supply would be in better shape. Geothermal plants, which are powered by the Earth's internal heat, can run for some time without human intervention.

According to the maintenance manual for the Svartsengi Island geothermal plant in Iceland, every six months the operators must change the gearbox oil and regrease all electric motors and couplings. Without humans to perform these sorts of maintenance procedures, some plants might run for a few years, but they'd all succumb to corrosion eventually.

Wind turbines

People relying on wind power would be in better shape than most. Turbines are designed so that they don't need constant maintenance, for the simple reason that there are a lot of them and they're a pain to climb.

Some windmills can run for a long time without human intervention. The Gedser Wind Turbine in Denmark was installed in the late 1950s, and generated power for 11 years without maintenance. Modern turbines are typically rated to run for 30,000 hours (three years) without servicing, and there are no doubt some that would run for decades. One of them would no doubt have at least a status LED in it somewhere.

Eventually, most of the wind turbines would be stopped by the same thing that would destroy the geothermal plants: Their gearboxes would seize up.

Hydroelectric dams

Generators that convert falling water into electricity will keep working for quite a while. The History Channel show *Life After People* spoke with an operator at the Hoover Dam, who said that if everyone walked out, the facility would continue to run on autopilot for several years. The dam would probably succumb to either clogged intakes or the same kind of mechanical failure that would hit the wind turbines and geothermal plants.

Batteries

Battery-powered lights will all be off in a decade or two. Even without anything using their power, batteries gradually self-discharge. Some types last longer than others, but even batteries advertised as having long shelf lives typically hold their charge only for a decade or two.

There are a few exceptions. In the Clarendon Laboratory at Oxford University sits a battery-powered bell that has been ringing since the year 1840. The bell "rings" so quietly it's almost inaudible, using only a tiny amount of charge with every motion of the clapper. Nobody knows exactly what kind of batteries it uses because nobody wants to take it apart to figure it out.

CERN PHYSICISTS INVESTIGATE THE OXFORD BELL

Sadly, there's no light hooked up to it.

Nuclear reactors

Nuclear reactors are a little tricky. If they settle into low-power mode, they can continue running almost indefinitely; the energy density of their fuel is just that high. As a certain webcomic put it:

Unfortunately, although there's enough fuel, the reactors wouldn't keep running for long. As soon as something went wrong, the core would go into automatic shutdown. This would happen quickly; many things can trigger it, but the most likely culprit would be a loss of external power.

It may seem strange that a power plant would require external power to run, but every part of a nuclear reactor's control system is designed so that a failure causes it to rapidly shut down, or "SCRAM."[1] When outside power is lost, either because the outside power plant shuts down or the on-site backup generators run out of fuel, the reactor would SCRAM.

---

[1] When Enrico Fermi built the first nuclear reactor, he suspended the control rods from a rope tied to a balcony railing. In case something went wrong, next to the railing was stationed a distinguished physicist with an axe. This led to the probably apocryphal story that SCRAM stands for "Safety Control Rod Axe Man."

Space probes

Out of all human artifacts, our spacecraft might be the longest-lasting. Some of their orbits will last for millions of years, although their electrical power typically won't.

Within centuries, our Mars rovers will be buried by dust. By then, many of our satellites will have fallen back to Earth as their orbits decayed. GPS satellites, in distant orbits, will last longer, but in time, even the most stable orbits will be disrupted by the Moon and Sun.

Many spacecraft are powered by solar panels, and others by radioactive decay. The Mars rover *Curiosity*, for example, is powered by the heat from a chunk of plutonium it carries in a container on the end of a stick.

MAGIC BOX OF DEATH

Curiosity could continue receiving electrical power from the RTG for over a century. Eventually the voltage will drop too low to keep the rover operating, but other parts will probably wear out before that happens.

So *Curiosity* looks promising. There's one problem: no lights.

Curiosity has lights; it uses them to illuminate samples and perform spectroscopy. However, these lights are turned on only when it's taking measurements. With no human instructions, it will have no reason to turn them on.

Unless they have humans on board, spacecraft don't need a lot of lights. The *Galileo* probe, which explored Jupiter in the 1990s, had several LEDs in the mechanism of its flight data recorder. Since they emitted infrared rather than visible light, calling them "lights" is a stretch—and in any case, *Galileo* was deliberately crashed into Jupiter in 2003.[2]

Other satellites carry LEDs. Some GPS satellites use, for example, UV LEDs to control charge buildup in some of their equipment, and they're powered by

---

2 The purpose of the crash was to safely incinerate the probe so it wouldn't accidentally contaminate the nearby moons, such as the watery Europa, with Earth bacteria.

solar panels; in theory they can keep running as long as the Sun is shining. Unfortunately, most won't even last as long as *Curiosity*; eventually, they'll succumb to space debris impacts.

But solar panels aren't used just in space.

Solar power

Emergency call boxes, often found along the side of the road in remote locations, are frequently solar-powered. They usually have lights on them, which provide illumination every night.

Like wind turbines, they're hard to service, so they're built to last for a long time. As long as they're kept free of dust and debris, solar panels will generally last as long as the electronics connected to them.

A solar panel's wires and circuits will eventually succumb to corrosion, but solar panels in a dry place, with well-built electronics, could easily continue providing power for a century if they're kept free of dust by occasional breezes or rain on the exposed panels.

If we follow a strict definition of lighting, solar-powered lights in remote locations could conceivably be the last surviving human light source.[3]

But there's another contender, and it's a weird one.

Cherenkov radiation

Radioactivity isn't usually visible.

MY WATCH DOESN'T GLOW ANYMORE.
TIME MARCHES ON; EVEN RADIUM'S FIRE CAN'T—
(THAT'S A CALCULATOR WATCH FROM
1991. THE BATTERY IS JUST DEAD.
...STILL, THOUGH. TIME.)

Watch dials used to be coated in radium, which made them glow. However, this glow didn't come from the radioactivity itself. It came from the phosphorescent paint on top of the radium, which glowed when it was irradiated. Over the years, the paint has broken down. Although the watch dials are still radioactive, they no longer glow.

Watch dials, however, are not our only radioactive light source.

When radioactive particles travel through materials like water or glass, they can emit light through a sort of optical sonic boom. This light is called Cherenkov radiation, and it's seen in the distinctive blue glow of nuclear reactor cores.

---

3 The USSR built some lighthouses powered by radioactive decay, but none are still in operation.

Some of our radioactive waste products, such as cesium-137, are melted and mixed with glass, then cooled into a solid block that can be wrapped in more shielding so they can be safely transported and stored.

In the dark, these glass blocks glow blue.

Cesium-137 has a half-life of thirty years, which means that two centuries later, they'll still be glowing with 1 percent of their original radioactivity. Since the color of the light depends only on the decay energy, and not the amount of radiation, it will fade in brightness over time but keep the same blue color.

And thus, we arrive at our answer: Centuries from now, deep in concrete vaults, the light from our most toxic waste will still be shining.

Q. Is it possible to build a jetpack using downward-firing machine guns?

—Rob B

- -

A. I WAS SORT OF surprised to find that the answer was yes! But to really do it right, you'll want to talk to the Russians.

The principle here is pretty simple. If you fire a bullet forward, the recoil pushes you back. So if you fire downward, the recoil should push you up.

The first question we have to answer is "can a gun even lift its own weight?" If a machine gun weighs 10 pounds but produces only 8 pounds of recoil when firing, it won't be able to lift itself off the ground, let alone lift itself plus a person.

In the engineering world, the ratio between a craft's thrust and the weight is called, appropriately, **thrust-to-weight ratio.** If it's less than 1, the vehicle can't lift off. The *Saturn V* had a takeoff thrust-to-weight ratio of about 1.5.

Despite growing up in the South, I'm not really a firearms expert, so to help answer this question, I got in touch with an acquaintance in Texas.[1]

Note: Please, PLEASE do not try this at home.

As it turns out, the AK-47 has a thrust-to-weight ratio of around 2. This means if you stood it on end and somehow taped down the trigger, it would rise into the air while firing.

---

1 Judging by the amount of ammunition they had lying around their house ready to measure and weigh for me, Texas has apparently become some kind of Mad Max–esque post-apocalyptic war zone.

SATURN V   KALASHNIKOV XLVII

This isn't true of all machine guns. The M60, for example, probably can't produce enough recoil to lift itself off the ground.

The amount of thrust created by a rocket (or firing machine gun) depends on (1) how much mass it's throwing out behind it, and (2) how fast it's throwing it. Thrust is the product of these two amounts:

$$\text{Thrust} = \text{Mass ejection rate} \times \text{Speed of ejection}$$

If an AK-47 fires ten 8-gram bullets per second at 715 meters per second, its thrust is:

$$10\,\tfrac{\text{bullets}}{\text{second}} \times 8\,\tfrac{\text{grams}}{\text{bullet}} \times 715\,\tfrac{\text{meters}}{\text{second}} = 57.2\text{N} \approx 13 \text{ pounds of force}$$

Since the AK-47 weighs only 10.5 pounds when loaded, it should be able to take off and accelerate upward.

In practice, the actual thrust would turn out to be up to around 30 percent higher. The reason for this is that the gun isn't spitting out just bullets—it's also spitting out hot gas and explosive debris. The amount of extra force this adds varies by gun and cartridge.

The overall efficiency also depends on whether you eject the shell casings out of the vehicle or carry them with you. I asked my Texan acquaintances if they could weigh some shell casings for my calculations. When they had trouble finding a scale, I helpfully suggested that given the size of their arsenal, really they just need to find someone *else* who owned a scale.[2]

So what does all this mean for our jetpack?

Well, the AK-47 could take off, but it doesn't have enough spare thrust to lift anything weighing much more than a squirrel.

We can try using multiple guns. If you fire two guns at the ground, it creates twice the thrust. If each gun can lift 5 pounds more than its own weight, two can lift 10.

---

2 Ideally someone with less ammo.

At this point, it's clear where we're headed:

You will not go to space today.

If we add enough rifles, the weight of the passenger becomes irrelevant; it's spread over so many guns that each one barely notices. As the number of rifles increases, since the contraption is effectively many individual rifles flying in parallel, the craft's thrust-to-weight ratio approaches that of a single, unburdened rifle:

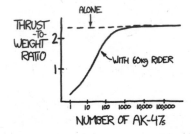

But there's a problem: ammunition.

An AK-47 magazine holds 30 rounds. At 10 rounds per second, this would provide a measly three seconds of acceleration.

We can improve this with a larger magazine—but only up to a point. It turns out there's no advantage to carrying more than about 250 rounds of ammunition. The reason for this is a fundamental and central problem in rocket science: Fuel makes you heavier.

Each bullet weighs 8 grams, and the cartridge (the "whole bullet") weighs over 16 grams. If we added more than about 250 rounds, the AK-47 would be too heavy to take off.

This suggests our optimal craft would comprise a large number of AK-47s (a minimum of 25 but ideally at least 300) carrying 250 rounds of ammunition each. The largest versions of this craft could accelerate upward to vertical speeds approaching 100 meters per second, climbing over half a kilometer into the air.

So we've answered Rob's question. With enough machine guns, you could fly.

But our AK-47 rig is clearly not a practical jetpack. Can we do better?

My Texas friends suggested a series of machine guns, and I ran the numbers on each one. Some did pretty well; the MG-42, a heavier machine gun, had a marginally higher thrust-to-weight ratio than the AK-47.

Then we went bigger.

The GAU-8 Avenger fires up to 60 1-pound bullets a *second*. It produces almost 5 tons of recoil force, which is crazy considering that it's mounted in a type of plane (the A-10 "Warthog") whose two engines produce only 4 tons of thrust each. If you put two of them in one aircraft, and fired both guns forward while opening up the throttle, the guns would win and you'd accelerate backward.

To put it another way: If I mounted a GAU-8 on my car, put the car in neutral, and started firing backward from a standstill, I would be breaking the interstate speed limit in less than *three seconds*.

"Actually, what I'm confused about is how."

As good as this gun would be as a rocket pack engine, the Russians built one that would work even better. The Gryazev-Shipunov GSh-6-30 weighs half as much as the GAU-8 and has an even higher fire rate. Its thrust-to-weight ratio approaches 40, which means if you pointed one at the ground and fired, not only would it take off in a rapidly expanding spray of deadly metal fragments, but you would experience 40 gees of acceleration.

This is way too much. In fact, even when it was firmly mounted in an aircraft, the acceleration was a problem:

[T]he recoil . . . still had a tendency to inflict damage on the aircraft. The rate of fire was reduced to 4,000 rounds a minute but it didn't help much. Landing lights almost always broke after firing . . . Firing more than about 30 rounds in a burst was asking for trouble from overheating . . .

— Greg Goebel, airvectors.net

But if you somehow braced the human rider, made the craft strong enough to survive the acceleration, wrapped the GSh-6-30 in an aerodynamic shell, and made sure it was adequately cooled . . .

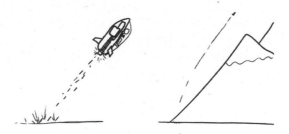

. . . you could jump mountains.

RISING STEADILY

Q. If you suddenly began rising steadily at 1 foot per second, how exactly would you die? Would you freeze or suffocate first? Or something else?

—Rebecca B

- -

A. DID YOU BRING A COAT?

A foot per second isn't that fast; it's substantially slower than a typical elevator. It would take you 5-7 seconds to rise out of arm's reach, depending how tall your friends are.

After 30 seconds, you'd be 30 feet—9 meters —off the ground. If you skip ahead to page 168, you'll learn that this is your last chance for a friend to throw you a sandwich or water bottle or something.[1]

After a minute or two you would be above the trees. For the most part, you'd still be about as comfortable as you were on the ground. If it's a breezy day, it would probably get chillier thanks to the steadier wind above the tree line.[2]

---

[1] It won't help you survive, but . . .

[2] For this answer, I'm going to assume a typical atmosphere temperature profile. It can, of course, vary quite a bit.

After 10 minutes you would be above all but the tallest skyscrapers, and after 25 minutes you'd pass the spire of the Empire State Building.

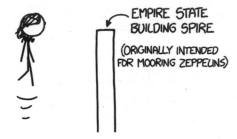

EMPIRE STATE
BUILDING SPIRE

(ORIGINALLY INTENDED
FOR MOORING ZEPPELINS)

The air at these heights is about 3 percent thinner than it is at the surface. Fortunately, your body handles air pressure changes like that all the time. Your ears might pop, but you wouldn't really notice anything else.

Air pressure changes quickly with height. Surprisingly, when you're standing on the ground, air pressure changes measurably within just a few feet. If your phone has a barometer in it, as a lot of modern phones do, you can download an app and actually see the pressure difference between your head and your feet.

A foot per second is pretty close to a kilometer per hour, so after an hour, you'll be about a kilometer off the ground. At this point, you definitely start to get chilly. If you have a coat, you'll still be OK, though you might also notice the wind picking up.

At about two hours and two kilometers, the temperature would drop below freezing. The wind would also, most likely, be picking up. If you have any exposed skin, this is where frostbite would start to become a concern.

At this point, the air pressure would fall below what you'd experience in an airliner cabin,[3] and the effects would start to become more significant. However, unless you had a warm coat, the temperature would be a bigger problem.

Over the next two hours, the air would drop to below-zero temperatures.[4,5] Assuming for a moment that you survived the oxygen deprivation, at some point you'd succumb to hypothermia. But when?

The scholarly authorities on freezing to death seem to be, unsurprisingly, Canadians. The most widely used model for human survival in cold air was developed by Peter Tikuisis and John Frim for the Defence and Civil Institute of Environmental Medicine in Ontario.

According to their model, the main factor in the cause of death would be your clothes. If you were nude, you'd probably succumb to hypothermia somewhere around the five-hour mark, before your oxygen ran out.[6] If you were bundled up, you may be frostbitten, but you would probably survive . . .

. . . long enough to reach the **Death Zone.**

Above 8000 meters—above the tops of all but the highest mountains—the oxygen content in the air is too low to support human life. Near this zone, you would experience a range of symptoms, possibly including confusion, dizziness, clumsiness, impaired vision, and nausea.

As you approach the Death Zone, your blood oxygen content would plummet. Your veins are supposed to bring low-oxygen blood back to your lungs to be

3 ...which are typically kept pressurized at about 70 percent to 80 percent of sea level pressure, judging from the barometer in my phone.
4 Either unit.
5 Not Kelvin, though.
6 And frankly, this "nude" scenario raises more questions than it answers.

refilled with oxygen. But in the Death Zone, there's so little oxygen in the air that your veins lose oxygen to the air instead of gaining it.

The result would be a rapid loss of consciousness and death. This would happen around the seven-hour mark; the chances are very slim that you would make it to eight.

She died as she lived—rising at a foot per second.
I mean, as she lived for the last few hours.

And two million years later, your frozen body, still moving along steadily at a foot per second, would pass through the heliopause into interstellar space.

Clyde Tombaugh, the astronomer who discovered Pluto, died in 1997. A portion of his remains were placed on the *New Horizons* spacecraft, which will fly past Pluto and then continue out of the solar system.

It's true that your hypothetical foot-per-second trip would be cold, unpleasant, and rapidly fatal. But when the Sun becomes a red giant in four billion years and consumes the Earth, you and Clyde would be the only ones to escape.

So there's that.

WEIRD (AND WORRYING) QUESTIONS FROM THE WHAT IF? INBOX, #3

Q. Given humanity's current knowledge and capabilities, is it possible to build a new star?

—Jeff Gordon

Q. What sort of logistic anomalies would you encounter in trying to raise an army of apes?

—Kevin

Q. If people had wheels and could fly, how would we differentiate them from airplanes?

—Anonymous

ORBITAL SUBMARINE

Q. How long could a nuclear submarine last in orbit?

—Jason Lathbury

- -

A. THE SUBMARINE WOULD BE fine, but the crew would be in trouble.

The submarine wouldn't burst. Submarine hulls are strong enough to withstand 50 to 80 atmospheres of external pressure from water, so they'd have no problem containing 1 atmosphere of internal pressure from air.

The hull would likely be airtight. Although watertight seals don't necessarily hold back air, the fact that water can't find a way through the hull under 50 atmospheres of pressure suggests that air won't escape quickly. There may be a few specialized one-way valves that would let air out, but in all likelihood, the submarine would remain sealed.

The big problem the crew would face would be the obvious one: air.

Nuclear submarines use electricity to extract oxygen from water. In space, there's no water,[*citation needed*] so they wouldn't be able to manufacture more air. They carry enough oxygen in reserve to survive for a few days, at least, but eventually they'd be in trouble.

To stay warm, they could run their reactor, but they'd have to be very careful how *much* they ran it—because the ocean is colder than space.

Technically, that's not really true. Everyone knows that space is very cold. The reason spacecraft can overheat is that space isn't as thermally conductive as water, so heat builds up more quickly in spacecraft than in boats.

But if you're even *more* pedantic, it *is* true. The ocean is colder than space.

Interstellar space is very cold, but space near the Sun—and near Earth—is actually incredibly hot! The reason it doesn't seem that way is that in space, the definition of "temperature" breaks down a little bit. Space seems cold because it's so *empty*.

Temperature is a measure of the average kinetic energy of a collection of particles. In space, individual molecules have a high average kinetic energy, but there are so few of them that they don't affect you.

When I was a kid, my dad had a machine shop in our basement, and I remember watching him use a metal grinder. Whenever metal touched the grinding wheel, sparks flew everywhere, showering his hands and clothes. I couldn't understand why they didn't hurt him—after all, the glowing sparks were several thousand degrees.

I later learned that the reason the sparks didn't hurt him was that they were *tiny*; the heat they carried could be absorbed into the body without warming anything more than a tiny patch of skin.

The hot molecules in space are like the sparks in my dad's machine shop; they might be hot or cold, but they're so small that touching them doesn't change your temperature much.[1] Instead, your heating and cooling is dominated by how much heat you produce and how quickly it pours out of you into the void.

Without a warm environment around you radiating heat back to you, you lose

---

[1] This is why, even though matches and torches are about the same temperature, you see tough guys in movies extinguish matches by pinching them but never see them do the same with torches.

heat by radiation much faster than normal. But without air around you to carry heat from your surface, you also don't lose much heat by convection.[2] For most human-carrying spacecraft, the latter effect is more important; the big problem isn't staying warm, it's keeping cool.

A nuclear submarine is clearly able to maintain a livable temperature inside when the outer hull is cooled to 4°C by the ocean. However, if the submarine's hull needed to hold this temperature while in space, it would lose heat at a rate of about 6 megawatts while in the shadow of the Earth. This is more than the 20 kilowatts supplied by the crew—and the few hundred kilowatts of apricity[3] when in direct sunlight—so they'd need to run the reactor just to stay warm.[4]

To get out of orbit, a submarine would need to slow down enough that it hit the atmosphere. Without rockets, it has no way to do this.

Okay—technically, a submarine *does* have rockets.

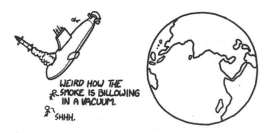

---

2 Or conduction.

3 This is my single favorite word in the English language. It means the warmth of sunlight in winter.

4 When they moved into the Sun, the sub's surface would warm, but they'd still be losing heat faster than they'd be gaining it.

Unfortunately, the rockets are pointing the wrong way to give the submarine a push. Rockets are self-propelling, which means they have very little recoil. When a gun fires a bullet, it's *pushing* the bullet up to speed. With a rocket, you just light it and let go. Launching missiles won't propel a submarine forward.

But *not* launching them could.

If the ballistic missiles carried by a modern nuclear submarine were taken from their tubes, turned around, and placed in the tubes backward, they could each change the submarine's speed by about 4 meters per second.

A typical de-orbiting maneuver requires in the neighborhood of 100 m/s of delta-v (speed change), which means that the 24 Trident missiles carried by an *Ohio*-class submarine could be just enough to get it out of orbit.

Now, because the submarine has no heat dissipating ablative tiles, and because it's not aerodynamically stable at hypersonic velocities, it would inevitably tumble and break up in the air.

If you tucked yourself into the right crevice in the submarine—and were strapped into an acceleration couch—there's a tiny, tiny, *tiny* chance that you could survive the rapid deceleration. Then you'd need to jump out of the wreckage with a parachute before it hit the ground.

If you ever try this, and I suggest you don't, I have one piece of advice that is absolutely critical:

Remember to disable the detonators on the missiles.

SHORT-ANSWER SECTION

Q. If my printer could literally print out money, would it have that big an effect on the world?

—Derek O'Brien

A. YOU CAN FIT FOUR bills on an 8.5" × 11" sheet of paper.

If your printer can manage one page (front and back) of full-color high-quality printing per minute, that's $200 million dollars a year.

This is enough to make you very rich, but not enough to put any kind of dent in the world economy. Since there are 7.8 billion $100 bills in circulation, and the lifetime of a $100 bill is about 90 months, that means there are about a billion produced each year. Your extra two million bills a year would barely be enough to notice.

LET'S SEE...
$400 PER MINUTE...

AND THERE ARE
♫ 525,600 MINUTES ♪
IN A YEAR...

(DAMMIT, RENT.)

Q. What would happen if you set off a nuclear bomb in the eye of a hurricane? Would the storm cell be immediately vaporized?

—Rupert Bainbridge (and hundreds of others)

A. THIS QUESTION GETS SUBMITTED a lot.

It turns out the National Oceanic and Atmospheric Administration—the agency that runs the National Hurricane Center—gets it a lot, too. In fact, they're asked about it so often that they've published a response.

I recommend you read the whole thing,[1] but I think the last sentence of the first paragraph says it all:

"Needless to say, this is not a good idea."

It makes me happy that an arm of the US government has, in some official capacity, issued an opinion on the subject of **firing nuclear missiles at hurricanes**.

Q. If everyone put little turbine generators on the downspouts of their houses and businesses, how much power would we generate? Would we ever generate enough power to offset the cost of the generators?

—Damien

---

[1] Search for "Why don't we try to destroy tropical cyclones by nuking them?" by Chris Landsea.

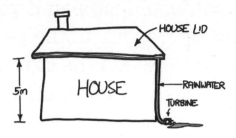

A. A HOUSE IN A very rainy place, like the Alaska panhandle, might receive close to 4 meters of rain per year. Water turbines can be pretty efficient. If the house has a footprint of 1500 square feet and gutters 5 meters off the ground, it would generate an average of less than a watt of power from rainfall, and the maximum electricity savings would be:

$$1500\text{ft}^2 \times 4\,\tfrac{\text{meters}}{\text{year}} \times 1\,\tfrac{\text{kg}}{\text{liter}} \times 9.81\,\tfrac{\text{m}}{\text{s}^2} \times 5\text{meters} \times 15\,\tfrac{\text{cents}}{\text{kWh}} = \frac{\$1.14}{\text{year}}$$

The rainiest hour on record as of 2014 occurred in 1947 in Holt, Missouri, where about 30 centimeters of rain fell in 42 minutes. For those 42 minutes, our hypothetical house could generate up to 800 watts of electricity, which might be enough to power everything inside it. For the rest of the year, it wouldn't come close.

If the generator rig cost $100, residents of the rainiest place in the US—Ketchikan, Alaska—could potentially offset the cost in under a century.

Q. Using only pronounceable letter combinations, how long would names have to be to give each star in the universe a unique one-word name?

—Seamus Johnson

A. THERE ARE ABOUT 300,000,000,000,000,000,000,000 stars in the universe. If you make a word pronounceable by alternating vowels and consonants (there are better ways to make pronounceable words, but this will do for an approximation), then every pair of letters you add lets you name 105 times as many stars (21 consonants times 5 vowels). Since numbers have a similar information density—100 possibilities per character—this suggests the name will end up being about as long as the total number of stars:

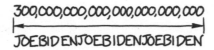

The stars are named Joe Biden.

I like doing math that involves measuring the lengths of numbers written out on the page (which is really just a way of loosely estimating $\log_{10} x$). It works, but it feels so *wrong*.

Q. I bike to class sometimes. It's annoying biking in the wintertime, because it's so cold. How fast would I have to bike for my skin to warm up the way a spacecraft heats up during reentry?

—David Nai

A. REENTERING SPACECRAFT HEAT UP because they're compressing the air in front of them (not, as is commonly believed, because of air friction).

To increase the temperature of the air layer in front of your body by 20 degrees Celsius (enough to go from freezing to room temperature), you would need to be biking at 200 meters per second.

The fastest human-powered vehicles at sea levels are recumbent bicycles enclosed in streamlined aerodynamic shells. These vehicles have an upper speed limit near 40 m/s—the speed at which the human can just barely produce enough thrust to balance the drag force from the air.

Since drag increases with the square of the speed, this limit would be pretty hard to push any further. Biking at 200 m/s would require at least 25 times the power output needed to go 40 m/s.

At those speeds, you don't really have to worry about the heating from the air—a quick back-of-the-envelope calculation suggests that if your body were doing that much work, your core temperature would reach fatal levels in a matter of seconds.

Q. How much physical space does the Internet take up?

—Max L

A. THERE ARE A LOT of ways to estimate the amount of information stored on the Internet, but we can put an interesting upper bound on the number just by looking at how much storage space we (as a species) have purchased.

The storage industry produces in the neighborhood of 650 million hard drives per year. If most of them are 3.5-inch drives, that's 8 liters (2 gallons) of hard drive per second.

This means the last few years of hard-drive production—which, thanks to increasing size, represents the majority of global storage capacity—would just about fill an oil tanker. So, by that measure, the Internet is smaller than an oil tanker.

Q. What if you strapped C4 to a boomerang? Could this be an effective weapon, or would it be as stupid as it sounds?

—Chad Macziewski

A. **AERODYNAMICS ASIDE, I'M CURIOUS** what tactical advantage you're expecting to gain by having the high explosive fly back at you if it misses the target.

LIGHTNING

Before we go any further, I want to emphasize something: **I am not an authority on lightning safety.**

I am a guy who draws pictures on the Internet. I like it when things catch fire and explode, which means I do not have your best interests in mind. The authorities on lightning safety are the folks at the US National Weather Service:

http://www.lightningsafety.noaa.gov/

Okay. With that out of the way . . .

To answer the questions that follow, we need to get an idea of where lightning is likely to go. There's a cool trick for this, and I'll give it away right here at the start: Roll an imaginary 60-meter sphere across the landscape and look at where it touches.[1] In this section, I answer a few different questions about lightning.

They say lightning strikes the tallest thing around. That's the kind of maddeningly inexact statement that immediately sparks all kinds of questions. How far is "around"? I mean, not all lightning hits Mount Everest. But does it find the tallest person in a crowd? The tallest person I know is probably Ryan North.[2] Should I try to hang around him for lightning safety reasons? What about other reasons? Maybe I should stick to answering questions rather than asking them.

So how *does* lightning pick its targets?

The strike starts with a branching bundle of charge—the "leader"—descending from the cloud. It spreads downward at speeds of tens to hundreds of kilometers per second, covering the few kilometers to the ground in a few dozen milliseconds.

The leader carries comparatively little current—on the order of 200 amps. That's still enough to kill you, but it's nothing compared to what happens next. Once the leader makes contact with the ground, the cloud and the ground equalize with a massive discharge of more like 20,000 amps. This is the blinding flash you see. It races back up the channel at a significant fraction of the speed of light, covering the distance in under a millisecond.[3]

---

1 Or a real one, for that matter.
2 Paleontologists estimate he stood nearly 5 meters tall at the shoulder.
3 While it's called a "return stroke," charge is still flowing downward. However, the discharge appears to propagate upward. This effect is similar to how when a traffic light turns green, the cars in front start moving, then the cars in back, so the movement appears to spread backward.

The place on the ground where we see a bolt "strike" is the spot where the leader first made contact with the surface. The leader moves down through the air in little jumps. It's ultimately making its way toward the (usually) positive charge in the ground. However, it "feels" charges within only a few tens of meters of its tip when it's deciding where to jump next. If there's something connected to the ground within that distance, the bolt will jump to it. Otherwise, it jumps out in a semi-random direction and repeats the process.

This is where the 60-meter sphere comes in. It's a way to imagine what spots might be the first thing the leader senses—the places it might jump to in its next (final) step.

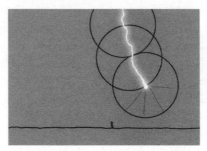

To figure out where lightning is likely to hit, you roll the imaginary 60-meter sphere across the landscape.[4] This sphere climbs up over trees and buildings without passing through anything (or rolling it up). Places the surface makes contact—treetops, fence posts, and golfers in fields—are potential lightning targets.

This means you can calculate a lightning "shadow" around an object of height h on a flat surface.

$$\text{Shadow Radius} = \sqrt{-h(h - 2r)}$$

The shadow is the area where the leader is likely to hit the tall object instead of the ground around it:

---

4 For safety reasons, do not use a real sphere.

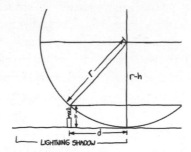

Now, that doesn't mean you're safe within the shadow—often, it means the opposite. After the current hits the tall object, it flows out into the ground. If you're touching the ground nearby, it can travel through your body. Of the 28 people killed by lightning in the US in 2012, 13 were standing under or near trees.

With all this in mind, let's look at possible lightning paths for the scenarios in the following questions.

Q. How dangerous is it, really, to be in a pool during a thunderstorm?

A. PRETTY DANGEROUS. WATER IS conductive, but that's not the biggest problem—the biggest problem is that if you're swimming, your head is poking up from a large flat surface. But lightning striking the water near you would still be bad. The 20,000 amps spread outward—mostly over the surface—but how much of a jolt it will give you at what distance is hard to calculate.

My guess is that you'd be in significant danger anywhere within a minimum of a dozen meters—and farther in fresh water, because the current will be happier to take a shortcut through you.

What would happen if you were taking a shower when you were struck by lightning? Or standing under a waterfall?

You're not in danger from the spray—it's just a bunch of droplets of water in the air. It's the tub under your feet, and the puddle of water in contact with the plumbing, that's the real threat.

Q. What would happen if you were in a boat or a plane that got hit by lightning? Or a submarine?

A. A BOAT WITHOUT A cabin is about as safe as a golf course. A boat with a closed cabin and a lightning protection system is about as safe as a car. A submarine is about as safe as a submarine safe (a submarine safe is not to be confused with a safe in a submarine—a safe in a submarine is substantially safer than a submarine safe).

Q. What if you were changing the light at the top of a radio tower, and lightning struck? Or what if you were doing a backflip? Or standing in a graphite field? Or looking straight up at the bolt?

 A.

Q. What would happen if lightning struck a bullet in midair?

A. THE BULLET WON'T AFFECT the path the lightning takes. You'd have to somehow time the shot so the bullet was in the middle of the bolt when the return stroke happened.

The core of a lightning bolt is a few centimeters in diameter. A bullet fired from an AK-47 is about 26 mm long and moves at about 700 millimeters every millisecond.

The bullet has a copper coating over a lead core. Copper is a fantastically good conductor of electricity, and much of the 20,000 amps could easily take a shortcut through the bullet.

Surprisingly, the bullet would handle it pretty well. If it were sitting still, the current would quickly heat and melt the metal. But since it would be moving along so quickly, it would exit the channel before it could be warmed by more than a few degrees. It would continue on to its target relatively unaffected. There would be some curious electromagnetic forces created by the magnetic field around the bolt and the current flow through the bullet, but none of the ones I examined would change the overall picture very much.

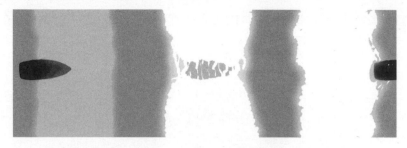

Q. What if you were flashing your BIOS during a thunderstorm and you got hit by lightning?

A.

WEIRD (AND WORRYING) QUESTIONS
FROM THE WHAT IF? INBOX, #4

Q. Would it be possible to stop a volcano eruption by placing a bomb (thermobaric or nuclear) underneath the surface?

—Tomasz Gruszka

Q. A friend of mine is convinced that there is sound in space. There isn't, right?

—Aaron Smith

Q. How much computing power could we could achieve if the entire world population stopped whatever we are doing right now and started doing calculations? How would it compare to a modern-day computer or smartphone?

—**Mateusz Knorps**

A. ON ONE HAND, HUMANS and computers do very different types of thinking, so comparing them is like comparing apples and oranges.

On the other hand, apples are better.[1] Let's try directly comparing humans and computers at the same tasks.

It's easy, though getting harder every day, to invent tasks that a single human can do faster than all the computers in the world. Humans, for example, are probably still far better at looking at a picture of a scene and guessing what just happened:

To test this theory, I sent this picture to my mother and asked her what *she* thought had happened. She immediately replied,[2] "The kid knocked over the vase and the cat is investigating."

She cleverly rejected alternate hypotheses, including:

- The cat knocked over the vase.
- The cat jumped out of the vase at the kid.
- The kid was being chased by the cat and tried to climb up the dresser with a rope to escape.
- There's a wild cat in the house, and someone threw a vase at it.
- The cat was mummified in the vase, but arose when the kid touched it with a magic rope.
- The rope holding the vase broke and the cat is trying to put it back together.
- The vase exploded, attracting a child and a cat. The child put on the hat for protection from future explosions.
- The kid and cat are running around trying to catch a snake. The kid finally caught it and tied a knot in it.

---

1   Except Red Delicious apples, whose misleading name is a travesty.
2   Our house had a lot of vases when I was a kid.

All the computers in the world couldn't figure out the correct answer faster than any one parent could. But that's because computers haven't been programmed to figure that kind of thing out,[3] whereas brains have been trained by millions of years of evolution to be good at figuring out what other brains around them are doing and why.

So we could choose a task to give the humans an advantage, but that's no fun; computers are limited by our ability to program them, so we've got a built-in advantage.

Instead, let's see how we compete on their turf.

The complexity of microchips

Rather than making up a new task, we'll simply apply the same benchmark tests to humans that we do to computers. These usually consist of things like floating point math, saving and recalling numbers, manipulating strings of letters, and basic logical calculations.

According to computer scientist Hans Moravec, a human running through computer chip benchmark calculations by hand, using pencil and paper, can carry out the equivalent of one full instruction every minute and a half.[4]

By this measure, the processor in a midrange mobile phone could do calculations about 70 times faster than the entire world population. A new high-end desktop PC chip would increase that ratio to 1500.

So, what year did a single typical desktop computer surpass the combined processing power of humanity?

---

3 Yet.
4 This figure comes from a list (*http://www.frc.ri.cmu.edu/users/hpm/book97/ch3/processor.list.txt*) in Hans Moravec's book *Robot: Mere Machine to Transcendent Mind*.

1994.

In 1992, the world population was 5.5 billion people, which means their combined computing power by our benchmark test was about 65 MIPS (million instructions per second).

That same year, Intel released the popular 486DX, which in its default configuration achieved about 55 or 60 MIPS. By 1994, Intel's new Pentium chips were achieving benchmark scores in the 70s and 80s, leaving humanity in the dust.

You might argue that we're being a little unfair to the computers. After all, these comparisons are one computer against all humans. How do all humans stack up against *all* computers?

This is tough to calculate. We can easily come up with benchmark scores for various types of computers, but how do you measure the instructions per second of, say, the chip in a Furby?

Most of the transistors in the world are in microchips not designed to run these tests. If we're assuming that all humans are being modified (trained) to carry out the benchmark calculations, how much effort should we spend to modify each computer chip so it can run the benchmark?

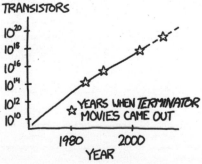

THE SQUARE ROOT OF 0.138338129 IS 0.37193834!

To avoid this problem, we can instead estimate the aggregate power of all the world's computing devices by counting transistors. It turns out that processors from the 1980s and processors from today have a roughly similar ratio of transistors to MIPS—about 30 transistors per instruction per second, give or take an order of magnitude.

A paper by Gordon Moore (of Moore's law fame) gives figures for the total number of transistors manufactured per year since the 1950s. It looks something like this:

Using our ratio, we can convert the number of transistors to a total amount of computing power. This tells us that a typical modern laptop, which has a benchmark score in the tens of thousands of MIPS, has more computing power than existed in the entire world in 1965. By that measure, the year when the combined power of computers finally pulled ahead of the combined computing power of humans was **1977**.

The complexity of neurons

Again, making people do pencil-and-paper CPU benchmarks is a *phenomenally* silly way to measure human computing power. Measured by complexity, our brains are more sophisticated than any supercomputer. Right?

Right. Mostly.

There are projects that attempt to use supercomputers to fully simulate a brain at the level of individual synapses.[5] If we look at how many processors and how much time these simulations require, we can come up with a figure for the number of transistors required to equal the complexity of the human brain.

The numbers from a 2013 run of the Japanese **K** supercomputer suggest a figure of 10^{15} transistors per human brain.[6] By this measure, it wasn't until the year 1988 that all the logic circuits in the world added up to the complexity of a single brain . . . and the total complexity of all our circuits is still dwarfed by the total complexity of all brains. Under Moore's law–based projections, and using these simulation figures, computers won't pull ahead of humans until the year **2036**.[7]

Why this is ridiculous

These two ways of benchmarking the brain represent opposite ends of a spectrum.

One, the pencil-and-paper Dhrystone benchmark, asks **humans** to manually simulate individual operations on a **computer** chip, and finds humans perform about 0.01 MIPS.

The other, the supercomputer neuron simulation project, asks **computers** to simulate individual neurons firing in a **human** brain, and finds humans perform about the equivalent of 50,000,000,000 MIPS.

---

5   Although even this might not capture everything that's going on. Biology is tricky.

6   Using 82,944 processors with about 750 million transistors each, **K** spent 40 minutes simulating one second of brain activity in a brain with 1 percent of the number of connections as a human's.

7   If it's past the year 2036 right now while you're reading this, hello from the distant past! I hope things are better in the future. P.S. Please figure out a way to come get us.

A slightly better approach might be to combine the two estimates. This actually makes a strange sort of sense. If we assume our computer programs are about as inefficient at simulating human brain activity as human brains are at simulating computer chip activity, then maybe a more fair brain power rating would be the geometric mean of the two numbers.

WAIT. I'M PRETTY SURE NOTHING IN THAT LAST SENTENCE WAS IN ANY WAY RIGOROUS.

The combined figure suggests human brains clock in at about 30,000 MIPS—right about on par with the computer on which I'm typing these words. It also suggests that the year when Earth's digital complexity overtook its human neurological complexity was **2004.**

Ants

In his paper "Moore's Law at 40," Gordon Moore makes an interesting observation. He points out that, according to biologist E. O. Wilson, there are 10^{15} to 10^{16} ants in the world. By comparison, in 2014 there were about 10^{20} transistors in the world, or tens of thousands of transistors per ant.[8]

An ant's brain might contain a quarter of a million neurons, and thousands of synapses per neuron, which suggests that the world's ant brains have a combined complexity similar to that of the world's human brains.

So we shouldn't worry too much about when computers will catch up with us in complexity. After all, we've caught up to ants, and *they* don't seem too concerned. Sure, we seem like we've taken over the planet, but if I had to bet on which one of us would still be around in a million years—primates, computers, or ants—I know who I'd pick.

Q. If an asteroid was very small but supermassive, could you really live on it like the Little Prince?

—Samantha Harper

"Did you eat my rose?" "Maybe."

A. THE LITTLE PRINCE, BY Antoine de Saint-Exupéry, is a story about a traveler from a distant asteroid. It's simple and sad and poignant and memorable.[1] It's ostensibly a children's book, but it's hard to pin down who the intended audience is. In any case, it certainly *has* found an audience; it's among the best-selling books in history.

---

1   Although not everyone sees it this way. Mallory Ortberg, writing on the-toast.net, characterized the story of *The Little Prince* as a wealthy child demanding that a plane crash survivor draw him pictures, then critiquing his drawing style.

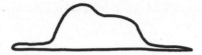

It was written in 1942. That's an interesting time to write about asteroids, because in 1942 we didn't actually know what asteroids *looked* like. Even in our best telescopes, the largest asteroids were visible only as points of light. In fact, that's where their name comes from — the word *asteroid* means "starlike."

We got our first confirmation of what asteroids looked like in 1971, when *Mariner 9* visited Mars and snapped pictures of Phobos and Deimos. These moons, believed to be captured asteroids, solidified the modern image of asteroids as cratered potatoes.

MARINER 9
IMAGE OF PHOBOS

Before the 1970s, it was common for science fiction to assume small asteroids would be round, like planets.

The Little Prince took this a step further, imagining an asteroid as a tiny planet with gravity, air, and a rose. There's no point in trying to critique the science here, because (1) it's not a story about asteroids, and (2) it opens with a parable about how foolish adults are for looking at everything too literally.

Rather than using science to chip away at the story, let's see what strange new pieces it can add. If there really were a superdense asteroid with enough surface gravity to walk around on, it would have some pretty remarkable properties.

If the asteroid had a radius of 1.75 meters, then in order to have Earthlike gravity at the surface, it would need to have a mass of about 500 million tons. This is roughly equal to the combined mass of every human on Earth.

If you stood on the surface, you'd experience tidal forces. Your feet would feel heavier than your head, which you'd feel as a gentle stretching sensation. It would feel like you were stretched out on a curved rubber ball, or were lying on a merry-go-round with your head near the center.

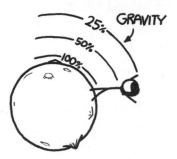

The escape velocity at the surface would be about 5 meters per second. That's slower than a sprint, but still pretty fast. As a rule of thumb, if you can't dunk a basketball, you wouldn't be able to escape this asteroid by jumping.

However, the weird thing about escape velocity is that it doesn't matter which direction you're going.[2] If you go faster than the escape speed, as long as you don't actually go *toward* the planet, you'll escape. That means you might be able to leave our asteroid by running horizontally and jumping off the end of a ramp.

---

2 …which is why it should really be called "escape speed"—the fact that it has no direction (which is the distinction between "speed" and "velocity") is unexpectedly significant here.

If you didn't go fast enough to escape the planet, you'd go into orbit around it. Your orbital speed would be roughly 3 meters per second, which is a typical jogging speed.

But this would be a *weird* orbit.

Tidal forces would act on you in several ways. If you stretched your arm down toward the planet, it would be pulled much harder than the rest of you. And when you reach down with one arm, the rest of you gets pushed upward, which means other parts of your body feel even *less* gravity. Effectively, every part of your body would be trying to go in a different orbit.

A large orbiting object under these kinds of tidal forces—say, a moon—will generally break apart into rings.[3] This wouldn't happen to you. However, your orbit would become chaotic and unstable.

These types of orbits were investigated in a paper by Radu D. Rugescu and Daniele Mortari. Their simulations showed that large, elongated objects follow strange paths around their central bodies. Even their centers of mass don't move in the traditional ellipses; some adopt pentagonal orbits, while others tumble chaotically and crash into the planet.

This type of analysis could actually have practical applications. There have

3   This is presumably what happened to Sonic the Hedgehog.

been various proposals over the years to use long, whirling tethers to move cargo in and out of gravity wells—a sort of free-floating space elevator. Such tethers could transport cargo to and from the surface of the Moon, or to pick up spacecraft from the edge of the Earth's atmosphere. The inherent instability of many tether orbits poses a challenge for such a project.

As for the residents of our superdense asteroid, they'd have to be careful; if they ran too fast, they'd be in serious danger of entering orbit, going into a tumble and losing their lunch.

Fortunately, vertical jumps would be fine.

Cleveland-area fans of French children's literature were disappointed by the Prince's decision to sign with the Miami Heat.

STEAK DROP

Q. From what height would you need to drop a steak for it to be cooked when it hit the ground?

—**Alex Lahey**

A. I HOPE YOU LIKE your steaks Pittsburgh Rare. And you may need to defrost it after you pick it up.

Things get really hot when they come back from space. As they enter the atmosphere, the air can't move out of the way fast enough, and gets squished in front of the object—and compressing air heats it up. As a rule of thumb, you start to notice compressive heating above about Mach 2 (which is why the Concorde had heat-resistant material on the leading edge of its wings).

When skydiver Felix Baumgartner jumped from 39 kilometers, he hit Mach 1 at around 30 kilometers. This was enough to heat the air by a few degrees, but the air was so far below freezing that it didn't make a difference. (Early in his jump, it was about minus 40 degrees, which is that magical point where you don't have to clarify whether you mean Fahrenheit or Celsius—it's the same in both.)

As far as I know, this steak question originally came up in a lengthy 4chan thread, which quickly disintegrated into poorly informed physics tirades intermixed with homophobic slurs. There was no clear conclusion.

To try to get a better answer, I decided to run a series of simulations of a steak falling from various heights.

An 8-ounce steak is about the size and shape of a hockey puck, so I based my steak's drag coefficients on those given on page 74 of *The Physics of Hockey* (which author Alain Haché actually measured personally using some lab equipment). A steak isn't a hockey puck, but the precise drag coefficient turned out not to make a big difference in the result.

Since answering these questions often includes analyzing unusual objects in extreme physical circumstances, often the only relevant research I can find is US military studies from the Cold War era. (Apparently, the US government was shoveling tons of money at anything even loosely related to weapons research.) To get an idea of how the air would heat the steak, I looked at research papers on the heating of ICBM nose cones as they reenter the atmosphere. Two of the most useful were "Predictions of Aerodynamic Heating on Tactical Missile Domes" and "Calculation of Reentry-Vehicle Temperature History."

Lastly, I had to figure out exactly how quickly heat spreads through a steak. I started by looking at some papers from industrial food production that simulated heat flow through various pieces of meat. It took me a while to realize there was a much easier way to learn what combinations of time and temperature will effectively heat the various layers of a steak: Check a cookbook.

Jeff Potter's excellent book *Cooking for Geeks* provides a great introduction to the science of cooking meat, and explains what ranges of heat produce what effects in steak and why. Cook's *The Science of Good Cooking* was also helpful.

Putting it all together, I found that the steak will accelerate quickly until it reaches an altitude of about 30–50 kilometers, at which point the air gets thick enough to start slowing it back down.

The falling steak's speed would steadily drop as the air gets thicker. No matter how fast it was going when it reached the lower layers of the atmosphere, it would quickly slow down to terminal velocity. No matter the starting height, it always takes six or seven minutes to drop from 25 kilometers to the ground.

For much of those 25 kilometers, the air temperature is below freezing—which means the steak will spend six or seven minutes subjected to a relentless blast of subzero, hurricane-force winds. Even if it's cooked by the fall, you'll probably have to defrost it when it lands.

When the steak does finally hit the ground, it will be traveling at terminal velocity—about 30 meters per second. To get an idea of what this means, imagine a steak flung at the ground by a major-league pitcher. If the steak is even partially frozen, it could easily shatter. However, if it lands in the water, mud, or leaves, it will probably be fine.[1]

A steak dropped from 39 kilometers will, unlike Felix, probably stay below the sound barrier. It also won't be appreciably heated. This makes sense—after all, Felix's suit wasn't scorched when he landed.

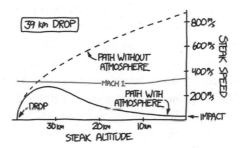

Steaks can probably survive breaking the sound barrier. In addition to Felix, pilots have ejected at supersonic speeds and lived to tell about it.

To break the sound barrier, you'll need to drop the steak from about 50 kilometers. But this still isn't enough to cook it.

We need to go higher.

If dropped from 70 kilometers, the steak will go fast enough to be briefly blasted by 350°F air. Unfortunately, this blast of thin, wispy air barely lasts a minute—and anyone with some basic kitchen experience can tell you that a steak placed in the oven at 350 for 60 seconds isn't going to be cooked.

From 100 kilometers—the formally defined edge of space—the picture's not much better. The steak spends a minute and a half over Mach 2, and the outer surface will likely be singed, but the heat is too quickly replaced by the icy stratospheric blast for it to actually be cooked.

At supersonic and hypersonic speeds, a shockwave forms around the steak that helps protect it from the faster and faster winds. The exact characteristics of this shock front—and thus the mechanical stress on the steak—depend on how an uncooked 8-ounce filet tumbles at hypersonic speeds. I searched the literature, but was unable to find any research on this.

For the sake of this simulation, I assume that at lower speeds some type of vortex shedding creates a flipping tumble, while at hypersonic speeds it's squished

---

1 I mean, intact. Not necessarily fine to *eat*.

into a semi-stable spheroid shape. However, this is little more than a wild guess. If anyone puts a steak in a hypersonic wind tunnel to get better data on this, *please*, send me the video.

If you drop the steak from 250 kilometers, things start to heat up; 250 kilometers puts us in the range of low Earth orbit. However, the steak, since it's dropped from a standstill, isn't moving nearly as fast as an object reentering from orbit.

In this scenario, the steak reaches a top speed of Mach 6, and the outer surface may even get pleasantly seared. The inside, unfortunately, is still uncooked. Unless, that is, it goes into a hypersonic tumble and explodes into chunks.

From higher altitudes, the heat starts to get really substantial. The shockwave in front of the steak reaches thousands of degrees (Fahrenheit or Celsius; it's true in both). The problem with this level of heat is that it burns the surface layer completely, converting it to little more than carbon. That is, it becomes charred.

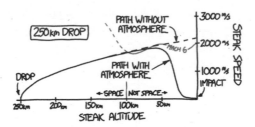

Charring is a normal consequence of dropping meat in a fire. The problem with charring meat at hypersonic speeds is that the charred layer doesn't have much structural integrity, and is blasted off by the wind—exposing a new layer to be charred. (If the heat is high enough, it will simply blast the surface layer off as it flash-cooks it. This is referred to in the ICBM papers as the "ablation zone.")

Even from those heights, the steak *still* doesn't spend enough time in the heat to get cooked all the way through.[2] We can try higher and higher speeds, and we might lengthen the exposure time via dropping it at an angle, from orbit.

But if the temperature is high enough or the burn time long enough, the steak will slowly disintegrate as the outer layer is repeatedly charred and blasted off. If most of the steak makes it to the ground, the inside will still be raw.

---

2 I know what some of you are probably thinking, and the answer is no—it doesn't spend enough time in the Van Allen belts to be sterilized by radiation.

STEAK DROP | 111

Which is why we should drop the steak over Pittsburgh.

As the probably apocryphal story goes, steelworkers in Pittsburgh would cook steaks by slapping them on the glowing metal surfaces coming out of the foundry, searing the outside while leaving the inside raw. This is, supposedly, the origin of the term "Pittsburgh Rare."

So drop your steak from a suborbital rocket, send out a collection team to recover it, brush it off, reheat it, cut away any badly charred sections, and dig in.

Just watch out for salmonella. And the Andromeda Strain.

Q. How hard would a puck have to be shot to be able to knock the goalie himself backward into the net?

—Tom

A. THIS CAN'T REALLY HAPPEN.

It's not just a problem of hitting the puck hard enough. This book isn't concerned with that kind of limitation. Humans with sticks can't make a puck go much faster than about 50 meters per second, but we can assume this puck is launched by a hockey robot or an electric sled or a hypersonic light gas gun.

The problem, in a nutshell, is that hockey players are heavy and pucks are not. A goalie in full gear outweighs a puck by a factor of about 600. Even the fastest slap shot has less momentum than a ten-year-old skating along at a mile per hour.

Hockey players can also brace pretty hard against the ice. A player skating at full speed can stop in the space of a few meters, which means the force they're exerting on the ice is pretty substantial. (It also suggests that if you started to slowly rotate a hockey rink, it could tilt up to 50 degrees before the players would all slide to one end. Clearly, experiments are needed to confirm this.)

From estimates of collision speeds in hockey videos, and some guidance from a hockey player, I estimated that the 165-gram puck would have to be moving somewhere between Mach 2 and Mach 8 to knock the goalie backward into the goal—faster if the goalie is bracing against the hit, and slower if the puck hits at an upward angle.

Firing an object at Mach 8 is not, in itself, very hard. One of the best methods for doing so is the aforementioned hypersonic gas gun, which is—at its core—the same mechanism a BB gun uses to fire BBs.[1]

But a hockey puck moving at Mach 8 would have a lot of problems, starting with the fact that the air ahead of the puck would be compressed and heated very rapidly. It wouldn't be going fast enough to ionize the air and leave a glowing trail like a meteor, but the surface of the puck would (given a long enough flight) start to melt or char.

The air resistance, however, would slow the puck down very quickly, so a puck going at Mach 8 when it leaves the launcher might be going a fraction of that when it arrives at the goal. And even at Mach 8, the puck probably wouldn't pass through the goalie's body. Instead, it would burst apart on impact with the power of a large firecracker or small stick of dynamite.

If you're like me, when you first saw this question, you might've imagined the puck leaving a cartoon-style hockey-puck-shaped hole. But that's because our intuitions are shaky about how materials react at very high speeds.

Instead, a different mental picture might be more accurate: Imagine throwing a ripe tomato—as hard as you can—at a cake.

That's about what would happen.

Q. If everyone on the planet stayed away from each other for a couple of weeks, wouldn't the common cold be wiped out?

— **Sarah Ewart**

A. WOULD IT BE WORTH IT?

The common cold is caused by a variety of viruses,[1] but rhinoviruses are the most common culprit.[2] These viruses take over the cells in your nose and throat and use them to produce more viruses. After a few days, your immune system notices and destroys it,[3] but not before you infect, on average, one other person.[4] After you fight off the infection, you are immune to that particular rhinovirus strain—an immunity that lasts for years.

If Sarah put us all in quarantine, the cold viruses we carry would have no fresh hosts to run to. Could our immune systems then wipe out every copy of the virus?

---

1 "Virii" is used occasionally but discouraged. "Viræ" is definitely wrong.
2 Any upper respiratory infection can actually be the cause of the "common cold."
3 The immune response is actually the cause of your symptoms, not the virus itself.
4 Mathematically, this must be true. If the average were less than one, the virus would die out. If it were more than one, eventually everyone would have a cold all the time.

Before we answer that question, let's consider the practical consequences of this kind of quarantine. The world's total annual economic output is in the neighborhood of $80 trillion, which suggests that interrupting all economic activity for a few weeks would cost many trillions of dollars. The shock to the system from the worldwide "pause" could easily cause a global economic collapse.

The world's total food reserves are probably large enough to cover us for four or five weeks of quarantine, but the food would have to be evenly parceled out beforehand. Frankly, I'm not sure what I'd do with a 20-day grain reserve while standing alone in a field somewhere.

A global quarantine brings us to another question: How far apart can we actually *get* from one another? The world is big,[citation needed] but there are a lot of people.[citation needed]

If we divide up the world's land area evenly, there's enough room for each of us to have a little over 2 hectares each, with the nearest person 77 meters away.

While 77 meters is probably enough separation to block the transmission of rhinoviruses, that separation would come at a cost. Much of the world's land is not pleasant to stand around on for five weeks. A lot of us would be stuck standing in the Sahara Desert,[5] or central Antarctica.[6]

A more practical—though not necessarily cheaper—solution would be to give everyone biohazard suits. That way, we could walk around and interact, even allowing some normal economic activity to continue:

But let's set aside the practicality and address Sarah's actual question: Would it *work*?

To help figure out the answer, I talked to Professor Ian M. Mackay, a virology expert from the Australian Infectious Diseases Research Centre at the University of Queensland.[7]

Dr. Mackay said that this idea is actually somewhat reasonable, from a purely biological point of view. He said that rhinoviruses—and other RNA respiratory viruses—are completely eliminated from the body by the immune system; they do not linger after infection. Furthermore, we don't seem to pass any rhinoviruses back and forth with animals, which means there are no other species that can serve as reservoirs of our colds. If rhinoviruses don't have enough humans to move between, they die out.

We've actually seen this viral extinction in action in isolated populations. The remote islands of St. Kilda, far to the northwest of Scotland, for centuries hosted a population of about 100 people. The islands were visited by only a few boats a year, and suffered from an unusual syndrome called the *cnatan-na-gall*, or

---

5 (450 million people).
6 (650 million people).
7 I first tried to take the question to *Boing Boing*'s Cory Doctorow, but he patiently explained to me that he's not actually a doctor.

"stranger's cough." For several centuries, the cough swept the island like clock-work every time a new boat arrived.

The exact cause of the outbreaks is unknown,[8] but rhinoviruses were probably responsible for many of them. Every time a boat visited, it would introduce new strains of virus. These strains would sweep the islands, infecting virtually every-one. After several weeks, all the residents would have fresh immunity to those strains, and with nowhere to go, the viruses would die out.

The same viral clearing would likely happen in any small and isolated popula-tion—for example, shipwreck survivors.

If all humans were isolated from one another, the St. Kilda scenario would play out on a species-wide scale. After a week or two, our colds would run their course, and healthy immune systems would have plenty of time to clear the viruses.

Unfortunately, there's one catch, and it's enough to unravel the whole plan: We don't all *have* healthy immune systems.

In most people, rhinoviruses are fully cleared from the body within about ten days. The story is different for those with severely weakened immune systems. In transplant pa-tients, for example, whose immune systems have been artificially suppressed, common in-

8   The residents of St. Kilda correctly identified the boats as the trigger for the outbreaks. The medical experts of the time, however, dismissed these claims, instead blaming the outbreaks on the way the islanders stood around outdoors in the cold when a boat arrived, and on their celebrating the new arrivals by drinking too much.

fections—including rhinoviruses—can linger for weeks, months, or conceivably years.

This small group of immunocompromised people would serve as safe havens for rhinoviruses. The hope of eradicating them is slim; they would need to survive in only a few hosts in order to sweep out and retake the world.

In addition to probably causing the collapse of civilization, Sarah's plan wouldn't eradicate rhinoviruses.[9] However, this might be for the best!

While colds are no fun, their absence might be worse. In his book *A Planet of Viruses,* author Carl Zimmer says that children who aren't exposed to rhinoviruses have more immune disorders as adults. It's possible that these mild infections serve to train and calibrate our immune systems.

On the other hand, colds suck. And in addition to being unpleasant, some research says infections by these viruses also *weaken* our immune systems directly and can open us up to further infections.

All in all, I wouldn't stand in the middle of a desert for five weeks to rid myself of colds forever. But if they ever come up with a rhinovirus vaccine, I'll be first in line.

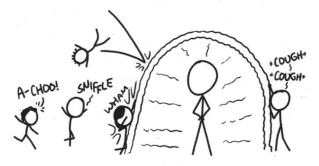

---

9 Unless we ran out of food during the quarantine and all starved to death; in that case, human rhinoviruses would die with us.

GLASS HALF EMPTY

Q. What if a glass of water was, all of a sudden, literally half empty?

—**Vittorio Iacovella**

- -

A. **THE PESSIMIST IS PROBABLY** more right about how it would turn out than the optimist.

When people say "glass half empty," they usually mean a glass containing equal parts water and air.

Traditionally, the optimist sees the glass as half full while the pessimist sees it as half empty. This has spawned a zillion joke variants—for example, the engineer sees a glass that's twice as big as it needs to be, the surrealist sees a giraffe eating a necktie, etc.

But what if the empty half of the glass were *actually* empty—a vacuum?[1] The

---

1 Even a vacuum arguably isn't truly empty, but that's a question for quantum semantics.

vacuum would definitely not last long. But exactly what happens depends on a key question that nobody usually bothers to ask: *Which* half is empty?

For our scenario, we'll imagine three different half-empty glasses, and follow what happens to them microsecond by microsecond.

In the middle is the traditional air/water glass. On the right is a glass like the traditional one, except the air is replaced by a vacuum. The glass on the left is half full of water and half empty—but it's the bottom half that's empty.

We'll imagine the vacuums appear at time **t=0**.

For the first handful of microseconds, nothing happens. On this timescale, even the air molecules are nearly stationary.

For the most part, air molecules jiggle around at speeds of a few hundred meters per second. But at any given time, some happen to be moving faster than others. The fastest few are moving at over 1000 meters per second. These are the first to drift into the vacuum in the glass on the right.

The vacuum on the left is surrounded by barriers, so air molecules can't easily get in. The water, being a liquid, doesn't expand to fill the vacuum in the same way air does. However, in the vacuum of the glasses, it does start to boil, slowly shedding water vapor into the empty space.

While the water on the surface in both glasses starts to boil away, in the glass on the right, the air rushing in stops it before it really gets going. The glass on the left continues to fill with a very faint mist of water vapor.

After a few hundred microseconds, the air rushing into the glass on the right fills the vacuum completely and rams into the surface of the water, sending a pressure wave through the liquid. The sides of the glass bulge slightly, but they contain the pressure and do not break. A shockwave reverberates through the water and back into the air, joining the turbulence already there.

The shockwave from the vacuum collapse takes about a millisecond to spread out through the other two glasses. The glass and water both flex slightly as the wave passes through them. In a few more milliseconds, it reaches the humans' ears as a loud bang.

Around this time, the glass on the left starts to visibly lift into the air.

The air pressure is trying to squeeze the glass and water together. This is the force we think of as suction. The vacuum on the right didn't last long enough for the suction to lift the glass, but since air can't get into the vacuum on the left, the glass and the water begin to slide toward each other.

The boiling water has filled the vacuum with a very small amount of water vapor. As the space gets smaller, the buildup of water vapor slowly increases the pressure on the water's surface. Eventually, this will slow the boiling, just like higher air pressure would.

However, the glass and water are now moving too fast for the vapor buildup to matter. Less than ten milliseconds after the clock started, they're flying toward each other at several meters per second. Without a cushion of air between them—only a few wisps of vapor—the water smacks into the bottom of the glass like a hammer.

Water is very nearly incompressible, so the impact isn't spread out over time—it comes as a single sharp shock. The momentary force on the glass is immense, and it breaks.

This "water hammer" effect (which is also responsible for the "clunk" you sometimes hear in old plumbing when you turn off the faucet) can be seen in the well-known party trick of smacking the top of a glass bottle to blow out the bottom.

When the bottle is struck, it's pushed suddenly downward. The liquid inside doesn't respond to the suction (air pressure) right away—much like in our scenario—and a gap briefly opens up. It's a small vacuum—a few fractions of an inch thick—but when it closes, the shock breaks the bottom of the bottle.

In our situation, the forces would be more than enough to destroy even the heaviest drinking glasses.

The bottom is carried downward by the water and thunks against the table. The water splashes around it, spraying droplets and glass shards in all directions.

Meanwhile, the detached upper portion of the glass continues to rise.

After half a second, the observers, hearing a pop, have begun to flinch. Their heads lift involuntarily to follow the rising movement of the glass.

The glass has just enough speed to bang against the ceiling, breaking into fragments . . .

. . . which, their momentum now spent, return to the table.

The lesson: If the optimist says the glass is half full, and the pessimist says the glass is half empty, the physicist ducks.

WEIRD (AND WORRYING) QUESTIONS
FROM THE WHAT IF? INBOX, #5

Q. If global warming puts us in danger through temperature rise, and super-volcanos put us into danger of global cooling, shouldn't those two dangers balance each other out?

—Florian Seidl-Schulz

Q. How fast would a human have to run in order to be cut in half at the bellybutton by a cheese-cutting wire?

—Jon Merrill

ALIEN ASTRONOMERS

Q. Let's assume there's life on the nearest habitable exoplanet and that they have technology comparable to ours. If they looked at our star right now, what would they see?

—**Chuck H**

A.

Let's try a more complete answer. We'll start with . . .

Radio transmissions

Contact popularized the idea of aliens listening in on our broadcast media. Sadly, the odds are against it.

Here's the problem: Space is really big.

You can work through the physics of interstellar radio attenuation,[1] but the problem is captured pretty well by considering the economics of the situation: If your TV signals are getting to another star, you're wasting money. Powering a transmitter is expensive, and creatures on other stars aren't buying the products in the TV commercials that pay your power bill.

The full picture is more complicated, but the bottom line is that as our technology has gotten better, less of our radio traffic has been leaking out into space. We're closing down the giant transmitting antennas and switching to cable, fiber, and tightly focused cell-tower networks.

While our TV signals may have been detectable—with great effort—for a while, that window is closing. Even in the late 20th century, when we were using TV and radio to scream into the void at the top of our lungs, the signal probably faded to undetectability after a few light-years. The potentially habitable exoplanets we've spotted so far are dozens of light-years away, so the odds are they aren't currently repeating our catchphrases.[2]

But TV and radio transmissions still weren't Earth's most powerful radio signal. They were outshone by the beams from **early-warning radar.**

Early-warning radar, a product of the Cold War, consisted of a bunch of ground and airborne stations scattered around the Arctic. These stations swept the atmosphere with powerful radar beams 24/7, often bouncing them off the ionosphere, and people obsessively monitored the echos for any hints of enemy movement.[3]

These radar transmissions leaked into space, and could probably be picked up by nearby exoplanets if they happened to be listening when the beam swept over their part of the sky. But the same march of technological progress that made the TV broadcast towers obsolete has had the same effect on early-warning radar.

---

1 I mean, if you want.
2 Contrary to the claims made by certain unreliable webcomics.
3 I wasn't alive during most of this period, but from what I hear, the mood was tense.

Today's systems—where they exist at all—are much quieter, and may eventually be replaced completely by new technology.

Earth's most *powerful* radio signal is the beam from the Arecibo telescope. This massive dish in Puerto Rico can function as a radar transmitter, bouncing a signal off nearby targets like Mercury and the asteroid belt. It's essentially a flashlight that we shine on planets to see them better. (This is just as crazy as it sounds.)

However, it transmits only occasionally, and in a narrow beam. If an exoplanet happened to be caught in the beam, and they were lucky enough to be pointing a receiving antenna at our corner of the sky at the time, all they would pick up would be a brief pulse of radio energy, then silence.[4]

So hypothetical aliens looking at Earth probably wouldn't pick us up with radio antennas.

But there's also . . .

Visible light

This is more promising. The Sun is really bright,[*citation needed*] and its light illuminates the Earth.[*citation needed*] Some of that light is reflected back into space as "Earthshine." Some of it skims close to our planet and passes through our atmo-

---

4 Which is exactly what we saw once, in 1977. The source of this blip (dubbed the "Wow Signal") has never been identified.

sphere before continuing on to the stars. Both of these effects could potentially be detected from an exoplanet.

They wouldn't tell you anything about humans directly, but if you watched the Earth for long enough, you could figure out a lot about our atmosphere from the reflectivity. You could probably figure out what our water cycle looked like, and our oxygen-rich atmosphere would give you a hint that something weird was going on.

So in the end, the clearest signal from Earth might not be from us at all. It might be from the algae that have been terraforming the planet—and altering the signals we send into space—for billions of years.

Heeeey, look at the time. Gotta run.

Of course, if we wanted to send a clearer signal, we could. A radio transmission has the problem that they have to be paying attention when it arrives.

Instead, we could *make* them pay attention. With ion drives, nuclear propulsion, or just clever use of the Sun's gravity well, we could probably send a probe out of the solar system fast enough to reach a given nearby star in a few dozen millennia. If we can figure out how to make a guidance system that survives the trip (which would be tough), we could use it to steer toward any inhabited planet.

To land safely, we'd have to slow down. But slowing down takes even more fuel. And, hey, the whole point of this was for them to notice us, right?

So maybe if those aliens looked toward our solar system, this is what they would see:

NO MORE DNA

Q. This may be a bit gruesome,
but . . . if someone's DNA
suddenly vanished, how long
would that person last?

—Nina Charest

A. IF YOU LOST YOUR DNA, you would instantly be about a third of a pound
lighter.

Losing a third of a pound

I don't recommend this strategy. There are easier ways to lose a third of a pound,
including:

- Taking off your shirt
- Peeing
- Cutting your hair (if you have very long hair)
- Donating blood, but putting a kink in the IV once they drain 150 mL and
  refusing to let them take any more
- Holding a 3-foot-diameter balloon full of helium
- Removing your fingers

You'll also lose a third of a pound if you take a trip from the polar regions to
the tropics. This happens for two reasons: One, the Earth is shaped like this:

EARTH

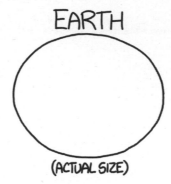

(ACTUAL SIZE)

If you stand on the North Pole, you're 20 kilometers closer to the center of the Earth than if you stand on the equator, and you feel a stronger pull from gravity.

Furthermore, if you're on the equator, you're being flung outward by centrifugal force.[1]

The result of these two phenomena is that if you move between polar regions and equatorial ones, you might lose or gain up to about half a percent of your body weight.

The reason I'm focusing on weight is that if your DNA disappeared, the physical loss of the matter wouldn't be the first thing you might notice. It's possible you'd feel something—a tiny, uniform shockwave as every cell contracted slightly—but maybe not.

If you were standing up when you lost your DNA, you might twitch slightly. When you stand, your muscles are constantly working to keep you upright. The force being exerted by those muscle fibers wouldn't change, but the mass they're

---

1 Yes, "centrifugal." I will fight you.

pulling on—your limbs—would. Since $F = ma$, various body parts would accelerate slightly.

After that, you would probably feel pretty normal.

For a while.

Destroying angel

Nobody has ever lost all their DNA,[2] so we can't say for sure what the precise sequence of medical consequences would be. But to get an idea of what it might be like, let's turn to mushroom poisonings.

Amanita bisporigera is a species of mushroom found in eastern North America. Along with related species in America and Europe, it's known by the common name **destroying angel.**

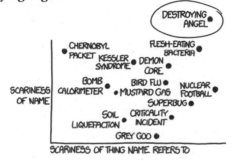

Destroying angel is a small, white, innocuous-looking mushroom. If you're like me, you were told never to eat mushrooms you found in the woods. *Amanita* is the reason why.[3]

If you eat a destroying angel, for the rest of the day you'll feel fine. Later that night, or the next morning, you'll start exhibiting cholera-like symptoms—vomiting, abdominal pain, and severe diarrhea. Then you start to feel better.

At the point where you start to feel better, the damage is probably irreversible. *Amanita* mushrooms contain **amatoxin,** which binds to an enzyme that is used to read information from DNA. It hobbles the enzyme, effectively interrupting the process by which cells follow DNA's instructions.

Amatoxin causes irreversible damage to whatever cells it collects in. Since

---

2 I don't have a citation for this, but I feel like we would have heard about it.
3 There are several members of the *Amanita* genus called "destroying angel," and—along with another *Amanita* called "death cap"—they are responsible for the vast majority of fatal mushroom poisonings.

most of your body is made of cells,[4] this is bad. Death is generally caused by liver or kidney failure, since those are the first sensitive organs in which the toxin accumulates. Sometimes intensive care and a liver transplant can be enough to save a patient, but a sizable percentage of those who eat *Amanita* mushrooms die.

The frightening thing about *Amanita* poisoning is the "walking ghost" phase—the period where you seem to be fine (or getting better), but your cells are accumulating irreversible and lethal damage.

This pattern is typical of DNA damage, and we'd likely see something like it in someone who lost their DNA.

The picture is even more vividly illustrated by two other examples of DNA damage: chemotherapy and radiation.

Radiation and chemotherapy

Chemotherapy drugs are blunt instruments. Some are more precisely targeted than others, but many simply interrupt cell division in general. The reason that this selectively kills cancer cells, instead of harming the patient and the cancer equally, is that cancer cells are dividing all the time, whereas most normal cells divide only occasionally.

Some human cells *do* divide constantly. The most rapidly dividing cells are found in the bone marrow, the factory that produces blood.

Bone marrow is also central to the human immune system. Without it, we lose the ability to produce white blood cells, and our immune system collapses. Chemotherapy causes damage to the immune system, which makes cancer patients vulnerable to stray infections.[5]

---

4 Citation: I got one of your friends to sneak into your room with a microscope while you were sleeping and check.

5 Immune boosters like pegfilgrastim (Neulasta) make frequent doses of chemotherapy safer. They stimulate white blood cell production by, in effect, tricking the body into thinking that it has a massive *E. coli* infection that it needs to fight off.

There are other types of rapidly dividing cells in the body. Our hair follicles and stomach lining also divide constantly, which is why chemotherapy can cause hair loss and nausea.

Doxorubicin, one of the most common and potent chemotherapy drugs, works by linking random segments of DNA to one another to tangle them. This is like dripping superglue on a ball of yarn; it binds the DNA into a useless tangle.[6] The initial side effects of doxorubicin, in the few days after treatment, are nausea, vomiting, and diarrhea—which makes sense, since the drug kills cells in the digestive tract.

A loss of DNA would cause similar cell death, and probably similar symptoms.

Radiation

Large doses of gamma radiation also harm you by damaging your DNA; radiation poisoning is probably the kind of real-life injury that most resembles Nina's scenario. The cells most sensitive to radiation are, as with chemotherapy, those in your bone marrow, followed by those in your digestive tract.[7]

Radiation poisoning, like destroying angel mushroom toxicity, has a latent period—a "walking ghost" phase. This is the period where the body is still working, but no new proteins can be synthesized and the immune system is collapsing.

In cases of severe radiation poisoning, the immune system collapse is the primary cause of death. Without a supply of white blood cells, the body can't fight off infections, and ordinary bacteria can get into the body and run wild.

The end result

Losing your DNA would most likely result in abdominal pain, nausea, dizziness, rapid immune system collapse, and death within days or hours from either rapid systemic infection or systemwide organ failure.

BUT I *LIKE* MY ORGANS!

On the other hand, there would be at least one silver lining. If we ever end up in a dystopian future where Orwellian governments collect our genetic information and use it to track and control us . . .

---

6 Although it's a little different; if you drip superglue on cotton thread, it will catch fire.
7 Extremely high radiation doses kill people quickly, but not because of DNA damage. Instead, they physically dissolve the blood-brain barrier, resulting in rapid death from cerebral hemorrhage (brain bleeding).

. . . you'd be invisible.

INTERPLANETARY CESSNA

Q. What would happen if you tried to fly a normal Earth airplane above different solar system bodies?

—Glen Chiacchieri

A. HERE'S OUR AIRCRAFT:[1]

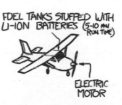

FUEL TANKS STUFFED WITH LI-ION BATTERIES (5-10 MIN RUN TIME)

ELECTRIC MOTOR

   We have to use an electric motor because gas engines work only near green plants. On worlds without plants, oxygen doesn't stay in the atmosphere—it combines with other elements to form things like carbon dioxide and rust. Plants undo this by stripping the oxygen back out and pumping it into the air. Engines need oxygen in the air to run.[2]

   Here's our pilot:

COME ON!

NOoooooo.

---

1 The Cessna 172 Skyhawk, probably the most common plane in the world.
2 Also, our gasoline is MADE of ancient plants.

Here's what would happen if our aircraft were launched above the surface of the 32 largest solar system bodies:

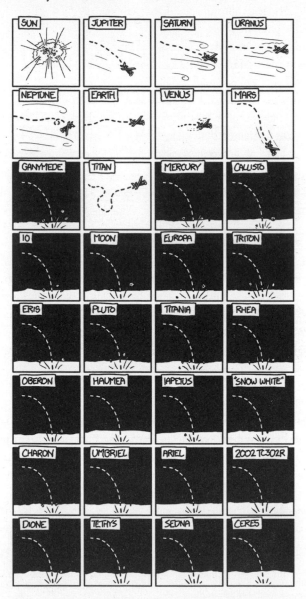

In most cases, there's no atmosphere, and the plane would fall straight to the ground. If it were dropped from 1 kilometer or less, in a few cases the crash would be slow enough that the pilot could survive—although the life-support equipment probably wouldn't.

There are nine solar system bodies with atmospheres thick enough to matter: Earth—obviously—Mars, Venus, the four gas giants, Saturn's moon Titan, and the Sun. Let's take a closer look at what would happen to a plane on each one.

The Sun: This would work about as well as you'd imagine. If the plane were released close enough to the Sun to feel its atmosphere at all, it would be vaporized in less than a second.

Mars: To see what would happen to our aircraft on Mars, we turn to X-Plane.

X-Plane is the most advanced flight simulator in the world. The product of 20 years of obsessive labor by a hardcore aeronautics enthusiast[3] and community of supporters, it actually simulates the flow of air over every piece of an aircraft's body as it flies. This makes it a valuable research tool, since it can accurately simulate entirely new aircraft designs—and new environments.

In particular, if you change the X-Plane config file to reduce gravity, thin the atmosphere, and shrink the radius of the planet, it can simulate flight on Mars.

X-Plane tells us that flight on Mars is difficult, but not impossible. NASA knows this, and has considered surveying Mars by airplane. The tricky thing is that with so little atmosphere, to get any lift, you have to go *fast*. You need to approach Mach 1 just to get off the ground, and once you get moving, you have so much inertia that it's hard to change course—if you turn, your plane rotates, but keeps moving in the original direction. The X-Plane author compared piloting Martian aircraft to flying a supersonic ocean liner.

Our Cessna 172 wouldn't be up to the challenge. If launched from 1 km, it wouldn't build up enough speed to pull out of a dive, and would plow into the Martian terrain at over 60 m/s (135 mph). If dropped from 4 or 5 kilometers, it could gain enough speed to pull up into a glide—at over half the speed of sound. The landing would not be survivable.

Venus: Unfortunately, X-Plane is not capable of simulating the hellish environment near the surface of Venus. But physics calculations give us an idea of what flight there would be like. The upshot is: Your plane would fly pretty well,

3 Who uses capslock a lot when talking about planes.

except it would be on fire the whole time, and then it would stop flying, and then stop being a plane.

The atmosphere on Venus is over 60 times denser than Earth's. It's thick enough that a Cessna moving at jogging speed would rise into the air. Unfortunately, that air is hot enough to melt lead. The paint would start melting off in seconds, the plane's components would fail rapidly, and the plane would glide gently into the ground as it came apart under the heat stress.

A much better bet would be to fly above the clouds. While Venus's surface is awful, its upper atmosphere is surprisingly Earthlike. At 55 kilometers, a human could survive with an oxygen mask and a protective wetsuit; the air is room temperature and the pressure is similar to that on Earth mountains. You would need the wetsuit, though, to protect you from the sulfuric acid.[4]

The acid's no fun, but it turns out the area right above the clouds is a great environment for an airplane, as long as it has no exposed metal to be corroded away by the sulfuric acid. And is capable of flight in constant category-5-hurricane-level winds, which are another thing I forgot to mention earlier.

Venus is a terrible place.

Jupiter: Our Cessna wouldn't be able to fly on Jupiter; the gravity is just too strong. The power needed to maintain level flight under Jupiter's gravity is three times greater than that on Earth. Starting from a friendly sea-level pressure, we'd accelerate through the tumbling winds into a 275 m/s (600 mph) downward glide deeper and deeper through the layers of ammonia ice and water ice until we and the aircraft were crushed. There's no surface to hit; Jupiter transitions smoothly from gas to liquid as you sink deeper and deeper.

Saturn: The picture here is a little friendlier than on Jupiter. The weaker gravity—close to Earth's, actually—and slightly denser (but still thin) atmosphere mean that we'd be able to struggle along a bit further before we gave in to either the cold or high winds and descended to the same fate as on Jupiter.

Uranus: Uranus is a strange, uniform bluish orb. There are high winds and it's bitterly cold. It's the friendliest of the gas giants to our Cessna, and you could probably fly for a little while. But given that it seems to be an almost completely featureless planet, why would you want to?

Neptune: If you're going to fly around one of the ice giants, I would probably

4 I'm not selling this well, am I?

recommend Neptune[5] over Uranus. It at least has some clouds to look at before you freeze to death or break apart from the turbulence.

Titan: We've saved the best for last. When it comes to flying, Titan might be better than Earth. Its atmosphere is thick but its gravity is light, giving it a surface pressure only 50 percent higher than Earth's with air four times as dense. Its gravity—lower than that of the Moon—means that flying is easy. Our Cessna could get into the air under pedal power.

In fact, humans on Titan could fly by muscle power. A human in a hang glider could comfortably take off and cruise around powered by oversized swim-flipper boots—or even take off by flapping artificial wings. The power requirements are minimal—it would probably take no more effort than walking.

The downside (there's always a downside) is the cold. It's 72 kelvin on Titan, which is about the temperature of liquid nitrogen. Judging from some numbers on heating requirements for light aircraft, I estimate that the cabin of a Cessna on Titan would probably cool by about 2 degrees per minute.

The batteries would help to keep themselves warm for a little while, but eventually the craft would run out of heat and crash. The Huygens probe, which descended with batteries nearly drained, taking fascinating pictures as it fell, succumbed to the cold after only a few hours on the surface. It had enough time to send back a single photo after landing—the only one we have from the surface of a body beyond Mars.

If humans put on artificial wings to fly, we might become Titanian versions of the Icarus story—our wings could freeze, fall apart, and send us tumbling to our deaths.

But I've never seen the Icarus story as a lesson about the limitations of humans. I see it as a lesson about the limitations of wax as an adhesive. The cold of Titan is just an engineering problem. With the right refitting, and the right heat sources, a Cessna 172 could fly on Titan—and so could we.

WEIRD (AND WORRYING) QUESTIONS
FROM THE WHAT IF? INBOX, #6

Q. What is the total nutritional value
(calories, fat, vitamins, minerals, etc.)
of the average human body?

—Justin Risner

Q. What temperature would a chainsaw
(or other cutting implement) need to be at to
instantly cauterize any injuries inflicted with it?

—Sylvia Gallagher

Q. How much Force power can Yoda output?

—Ryan Finnie

A. I'M GOING TO — of course — ignore the prequels.

Yoda's greatest display of raw power in the original trilogy came when he lifted Luke's X-wing from the swamp. As far as physically moving objects around goes, this was easily the biggest expenditure of energy through the Force we saw from anyone in the trilogy.

The energy it takes to lift an object to a given height is equal to the object's mass times the force of gravity times the height it's lifted. The X-wing scene lets us use this to put a lower limit on Yoda's peak power output.

First we need to know how heavy the ship was. The X-wing's mass has never been canonically established, but its length has — 12.5 meters. An F-22 is 19 meters long and weighs 19,700 kg, so scaling down from this gives an estimate for the X-wing of about 12,000 pounds (5 metric tons).

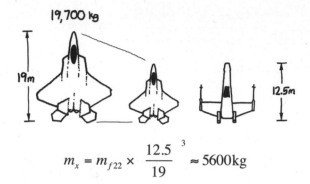

$$m_x = m_{f22} \times \frac{12.5}{19}^{\,3} \approx 5600\text{kg}$$

Next, we need to know how fast it was rising. I went over footage of the scene and timed the X-wing's rate of ascent as it was emerging from the water.

The front landing strut rises out of the water in about three and a half seconds, and I estimated the strut to be 1.4 meters long (based on a scene in *A New Hope* where a crew member squeezes past it), which tells us the X-wing was rising at 0.39 m/s.

Lastly, we need to know the strength of gravity on Dagobah. Here, I figure I'm stuck, because while sci-fi fans are obsessive, it's not like there's gonna be a catalog of minor geophysical characteristics for every planet visited in *Star Wars*. Right?

Nope. I've underestimated the fandom. Wookieepeedia has just such a catalog, and informs us that the surface gravity on Dagobah is 0.9g. Combining this with the X-wing mass and lift rate gives us our peak power output:

$$\frac{5600\text{kg} \times 0.9\text{g} \times 1.4 \text{ meters}}{3.6 \text{ seconds}} = 19.2\text{kW}$$

That's enough to power a block of suburban homes. It's also equal to about 25 horsepower, which is about the power of the motor in the electric-model Smart Car.

At current electricity prices, Yoda would be worth about $2/hour.

But telekinesis is just one type of Force power. What about that lightning the Emperor used to zap Luke? The physical nature of it is never made clear, but Tesla coils that produce similar displays draw something like 10 kilowatts—which would put the Emperor roughly on par with Yoda. (Those Tesla coils typically use lots of very short pulses. If the Emperor is sustaining a continuous arc, as in an arc welder, the power could easily be in the megawatts.)

What about Luke? I examined the scene where he used his nascent Force powers to yank his lightsaber out of the snow. The numbers are harder to estimate here, but I went through frame-by-frame and came up with an estimate of 400 watts for his peak output. This is a fraction of Yoda's 19kW, and was sustained for only a fraction of a second.

So Yoda sounds like our best bet as an energy source. But with world electricity consumption pushing 2 terawatts, it would take a hundred million Yodas to meet our demands. All things considered, switching to Yoda power probably isn't worth the trouble—though it would *definitely* be green.

Q. Which US state is actually flown over the most?

—Jesse Ruderman

A. WHEN PEOPLE SAY "FLYOVER states," they're usually referring to the big, square states out west that people stereotypically cross over while flying between New York, LA, and Chicago, but don't actually land in.

But what state do the largest number of planes *actually* fly over? There are a lot of flights up and down the East Coast; it would be easy to imagine that people fly over New York more often than Wyoming.

To figure out what the real flyover states are, I looked at over 10,000 air traffic routes, determining which states each flight passed over.

Surprisingly, the state with the most planes flying over it—without taking off or landing—is . . .

. . . **Virginia.**

This result surprised me. I grew up in Virginia, and I certainly never thought of it as a "flyover state."

It's surprising because Virginia has several major airports; two of the airports serving DC are actually located in Virginia (DCA/Reagan and IAD/Dulles). This means most flights to DC don't count toward flights over Virginia, since those flights *land* in Virginia.

Here's a map of US states colored by number of daily flyovers:

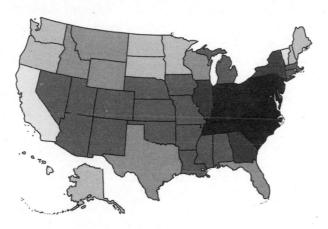

Close behind Virginia are **Maryland, North Carolina,** and **Pennsylvania.** These states have substantially more daily flyovers than any other.

So why Virginia?

There are a number of factors, but one of the biggest is **Hartsfield-Jackson Atlanta International Airport.**

Atlanta's airport is the busiest in the world, with more passengers and flights than Tokyo, London, Beijing, Chicago, or Los Angeles. It's the main hub airport for Delta Air Lines—until recently the world's largest airline—which means passengers taking Delta flights will often connect through Atlanta.

Thanks to the large volume of flights from Atlanta to the northeast US, 20 percent of all Atlanta flights cross Virginia and 25 percent cross North Carolina, contributing substantially to the totals for each state.

However, Atlanta isn't the biggest contributor to Virginia's totals. The airport with the most flights over Virginia was a surprise to me.

Toronto Pearson International Airport (YYZ) seems an unlikely source of Virginia-crossing flights, but Canada's largest airport contributes more flights over Virginia than New York's JFK and LaGuardia airports *combined*.

Part of the reason for Toronto's dominance is that it has many direct flights to the Caribbean and South America, which cross US airspace on the way to their destinations.[1] In addition to Virginia, Toronto is also the chief source of flights over West Virginia, Pennsylvania, and New York.

This map shows, for each state, which airport is the source of the most flights over it:

Flyover states by ratio

Another possible definition of "flyover state" is the state that has the highest ratio of flights *over* it to flights *to* it. By this measure, the flyover states are, for the most part, simply the least dense states. The top ten include, predictably, Wyoming, Alaska, Montana, Idaho, and the Daktoas.

The state with the *highest* ratio of flights-over-to-flights-to, however, is a surprise: **Delaware.**

A little digging turned up the very straightforward reason: Delaware has no airports.

Now, that's not quite true. Delaware has a number of airfields, including Dover Air Force Base (DOV) and New Castle Airport (ILG). New Castle Air-

---

[1] It helps that Canada, unlike the US, has extensive commercial flight service to Cuba.

port is the only one that might qualify as a commercial airport, but after Skybus Airlines shut down in 2008, the airport had no airlines serving it.[2]

Least flown-over state

The least flown-over state is Hawaii, which makes sense. It consists of tiny islands in the middle of the world's biggest ocean; you have to try pretty hard to hit it.

Of the 49 non-island states,[3] the least flown-over state is California. This came as a surprise to me, since California is long and skinny, and it seems like a lot of flights over the Pacific would need to pass over it.

However, since jet-fuel-laden planes were used as weapons on 9/11, the FAA has tried to limit the number of unnecessarily fuel-heavy flights crossing the US, so most international travelers who might otherwise travel over California instead take a connecting flight from one of the airports there.

Fly-under states

Lastly, let's answer a slightly stranger question: What is the most flown-*under* state? That is, what state has the most flights on the opposite side of the Earth pass directly under its territory?

The answer turns out to be **Hawaii.**

The reason such a tiny state wins in this category is that most of the US is opposite the Indian Ocean, which has very few commercial flights over it. Hawaii, on the other hand, is opposite Botswana in Central Africa. Africa doesn't have a high volume of flights over it compared to most other continents, but it's enough to win Hawaii the top spot.

Poor Virginia

As someone who grew up there, it's hard for me to accept Virginia's status as the most flown-over state. If nothing else, when I'm back home with family, I'll try to remember—once in a while—to look up and wave.

(And if you find yourself on Arik Air Flight 104 from Johannesburg, South Africa to Lagos, Nigeria—daily service, departing at 9:35 A.M.—remember to look down and say "Aloha!")

---

2 This changed in 2013, when Frontier Airlines began operating a route between New Castle Airport and Fort Myers, Florida. This wasn't included in my data set, and it's possible Frontier will bump Delaware down the list.
3 I'm including Rhode Island here, although it seems wrong to.

FALLING WITH HELIUM

Q. What if I jumped out of an airplane with a couple of tanks of helium and one huge, un-inflated balloon? Then, while falling, I release the helium and fill the balloon. How long of a fall would I need in order for the balloon to slow me enough that I could land safely?

—Colin Rowe

- -

Q. AS RIDICULOUS AS IT sounds, this is—sort of—plausible.

Falling from great heights is dangerous.[*citation needed*] A balloon could actually help save you, although a regular helium one from a party obviously won't do the trick.

If the balloon is large enough, you don't even need the helium. A balloon will act as a parachute, slowing your fall to nonfatal speeds.

Avoiding a high-speed landing is, unsurprisingly, the key to survival. As one medical paper put it . . .

> It is, of course, obvious that speed, or height of fall, is not in itself injurious . . . but a high rate of change of velocity, such as occurs after a 10 story fall onto concrete, is another matter.

. . . which is just a wordy version of the old saying "It's not the fall that kills you, it's the sudden stop at the end."

To act as a parachute, a balloon filled with air—rather than helium—would have to be 10 to 20 meters across, far too big to be inflated with portable tanks. A powerful fan could be used to fill it with ambient air, but at that point, you may as well just use a parachute.

Helium

The helium makes things easier.

It doesn't take too many helium balloons to lift a person. In 1982, Larry Walters flew across Los Angeles in a lawn chair lifted by weather balloons, eventually reaching several miles in altitude. After passing through LAX airspace, he descended by shooting some of the balloons with a pellet gun.

On landing, Walters was arrested, although the authorities had some trouble figuring out what to charge him with. At the time, an FAA safety inspector told the *New York Times*, "We know he broke some part of the Federal Aviation Act, and as soon as we decide which part it is, some type of charge will be filed."

A relatively small helium balloon—certainly smaller than a parachute—will suffice to slow your fall, but it still has to be huge by party balloon standards. The biggest consumer rental helium tanks are about 250 cubic feet, and you would need to empty at least ten of them to put enough air in the balloon to support your weight.

You'd have to do it quickly. Compressed helium cylinders are smooth and often quite heavy, which means they have a high terminal velocity. You'll have only a few minutes to use up all the cylinders. (As soon as you emptied one, you could drop it.)

You can't get around this problem by moving your starting point higher. As we learned from the steak incident, since the upper atmosphere is pretty thin, anything dropped from the stratosphere or higher will accelerate to very high speeds until it hits the lower atmosphere, then fall slowly the rest of the way. This is true of everything from small meteors[1] to Felix Baumgartner.

But if you inflated the balloons quickly, possibly by connecting many canisters to it at once, you'd be able to slow your fall. Just don't use too much helium, or you'll end up floating at 16,000 feet like Larry Walters.

While researching this answer, I managed to lock up my copy of Mathematica several times on balloon-related differential equations, and subsequently got my IP address banned from Wolfram|Alpha for making too many requests. The ban-appeal form asked me to explain what task I was performing that necessitated so many queries. Calculating how many rental helium tanks you'd have to carry with you in order to inflate a balloon large enough to act as parachute and slow your fall from a jet aircraft.

Sorry, Wolfram.

---

1 While researching impact speeds for this answer, I came across a discussion on the Straight Dope Message Board about survivable fall heights. One poster compared a fall from height to being hit by a bus. Another user, a medical examiner, replied that this was a bad comparison:

"When hit by a car, the vast majority of people are not run over; they are run under. The lower legs break, sending them into the air. They usually strike the hood of the car, often with the back of the head impacting the windshield, 'starring' the windshield, possibly leaving a few hairs in the glass. They then go over the top of the car. They are still alive, although with broken legs, and maybe with head pain from the nonfatal windshield impact. They die when they hit the ground. They die from head injury."

The lesson: Don't mess with medical examiners. They're apparently pretty hardcore.

EVERYBODY OUT

Q. Is there enough energy to move the entire current human population off-planet?

—Adam

A. THERE ARE A BUNCH of science fiction movies where, because of pollution, overpopulation, or nuclear war, humanity abandons Earth.

But lifting people into space is hard. Barring a massive reduction in the population, is launching the whole human race into space physically possible? Let's not even worry about where we're headed—we'll assume we don't have to find a new home, but we can't stay here.

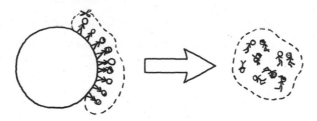

To figure out if this is plausible, we can start with an absolute baseline energy requirement: 4 gigajoules per person. No matter how we do it, whether we use rockets or a cannon or a space elevator or a ladder, moving a 65-kilogram person—or 65 kilograms of anything—out of the Earth's gravity well requires at least this much energy.

How much is 4 gigajoules? It's about a megawatt-hour, which is what a typical US household consumes in electricity in a month or two. It's equal to the amount of stored energy in 90 kg of gasoline or a cargo van full of AA batteries.

Four gigajoules times seven billion people gives us 2.8×10^{18} joules, or 8 petawatt-hours. This is about 5 percent of the world's annual energy consumption. A lot, but not physically implausible.

However, 4 gigajoules is just a minimum. In practice, everything would depend on our means of transportation. If we were using rockets, for example, it would take a lot more energy than that. This is because of a fundamental problem with rockets: They have to lift their own fuel.

Let's return for a moment to those 90 kilograms of gasoline (about 30 gallons), because they help illustrate this central problem in space travel.

If we want to launch a 65-kilogram spaceship, we need the energy of around 90 kilograms of fuel. We load that fuel on board—and now our spaceship weighs 155 kilograms. A 155-kilogram spaceship requires 215 kilograms of fuel, so we load another 125 kilograms on board . . .

Fortunately, we're saved from an infinite loop—where we add 1.3 kilograms for every 1 kilogram we add—by the fact that we don't have to carry that fuel all the way up. We burn it as we go, so we get lighter and lighter, which means we need less and less fuel. But we do have to lift the fuel partway. The formula for how much propellant we need to burn to get moving at a given speed is given by the Tsiolkovsky Rocket equation:

$$\Box_\ast \quad v_{exhaust} \ln \frac{m_{start}}{m_{end}}$$

m_{start} and m_{end} are the total mass of the ship plus the fuel before and after the burn, and $v_{exhaust}$ is the "exhaust velocity" of the fuel, a number that's between 2.5 and 4.5 km/s for rocket fuels.

What's important is the ratio between Δv, the speed we want to be going, and $v_{exhaust}$, the speed that the propellant exits our rocket. For leaving Earth, we need a Δv of upward of 13 km/s, and $v_{exhaust}$ is limited to about 4.5 km/s, which gives a fuel-to-ship ratio of at least $e^{\frac{13}{4.5}} \approx 20$. If that ratio is x, then to launch a kilogram of ship, we need e^x kilograms of fuel.

As x grows, this amount gets very large.

The upshot is that to overcome Earth's gravity using traditional rocket fuels, a 1-ton craft needs 20 to 50 tons of fuel. Launching all of humanity (total weight: around 400 million tons) would therefore take tens of trillions of tons of fuel. That's a lot; if we were using hydrocarbon-based fuels, it would represent a decent chunk of the world's remaining oil reserves. And that's not even worrying about the weight of the ship itself, food, water, or our pets.[1] We'd also need fuel to produce all these ships, to transport people to the launch sites, and so forth. It's not necessarily completely impossible, but it's certainly outside the realm of plausibility.

But rockets aren't our only option. As crazy as it sounds, we might be better off trying to (1) literally climb into space on a rope, or (2) blow ourselves off the planet with nuclear weapons. These are actually serious—if audacious—ideas for launch systems, both of which have been bouncing around since the start of the Space Age.

The first approach is the "space elevator" concept, a favorite of science fiction authors. The idea is that we connect a tether to a satellite orbiting far enough out that the tether is held taut by centrifugal force. Then we can send climbers up the rope using ordinary electricity and motors, powered by solar power, nuclear generators, or whatever works best. The biggest engineering hurdle is that the tether would have to be several times stronger than anything we can currently

---

[1] There are probably around a million tons of pet dog in the US alone.

build. There are hopes that carbon nanotube-based materials could provide the required strength—adding this to the long list of engineering problems that can be waved away by tacking on the prefix "nano-."

The second approach is nuclear pulse propulsion, a surprisingly plausible method for getting huge amounts of material moving really fast. The basic idea is that you toss a nuclear bomb behind you and ride the shockwave. You'd think the spacecraft would be vaporized, but it turns out that if it has a well-designed shield, the blast would fling away before it has a chance to disintegrate. If it could be made reliable enough, this system would in theory be capable of lifting entire city blocks into orbit, and could—potentially—accomplish our goal.

The engineering principles behind this were thought to be solid enough that in the 1960s, under the guidance of Freeman Dyson, the US government actually tried to build one of these spaceships. The story of that effort, dubbed **Project Orion,** is detailed in the excellent book of the same name by Freeman's son, George. Advocates for nuclear pulse propulsion are still disappointed that the project was cancelled before any prototypes were built. Others argue that when you think about what they were trying to do—put a gigantic nuclear arsenal in a box, hurl it high into the atmosphere, and bomb it repeatedly—it's terrifying that it got as far as it did.

So the answer is that while sending one person into space is easy, getting all of us there would tax our resources to the limit and possibly destroy the planet. It's a small step for a man, but a giant leap for mankind.

WEIRD (AND WORRYING) QUESTIONS
FROM THE WHAT IF? INBOX, #7

Q. In *Thor* the main character is at one point spinning his hammer so fast that he creates a strong tornado. Would this be possible in real life?

—**Davor**

NO.

Q. If you saved a whole life's worth of kissing and used all that suction power on one single kiss, how much suction force would that single kiss have?

—**Jonatan Lindström**

Q. How many nuclear missiles would have to be launched at the United States to turn it into a complete wasteland?

—**Anonymous**

Q. I read about some researchers who were trying to produce sperm from bone marrow stem cells. If a woman were to have sperm cells made from her own stem cells and impregnate herself, what would be her relationship to her daughter?

—R Scott LaMorte

A. TO MAKE A HUMAN, you need to put together two sets of DNA.

HOW DO YOU PUT THEM TOGETHER?

ASK YOUR PARENTS. OR THE INTERNET.

In humans, these two sets are held in a sperm cell and an egg cell, each of which holds a random sample of the parents' DNA. (More on how that randomization works in a moment.) In humans, these cells are from two different people.

However, that doesn't necessarily have to be the case. Stem cells, which can form any type of tissue, could in principle be used to produce sperm (or eggs).

So far, nobody has been able to produce complete sperm from stem cells. In 2007, a group of researchers succeeded in turning bone marrow stem cells into spermatogonial stem cells. These cells are the predecessors to sperm. The researchers couldn't get the cells to fully develop into sperm, but it was a step. In 2009, the same group published a paper that seemed to claim they'd made the final step and produced functioning sperm cells.

There were two problems.

First, they didn't actually *say* they had produced sperm cells. They said they produced sperm-*like* cells, but the media generally glossed over this. Second, the paper was retracted by the journal that published it. It turns out the authors had plagiarized two paragraphs of their article from another paper.

Despite these problems, the fundamental idea here is not that far-fetched, and the answer to R. Scott's question turns out to be a little bit unsettling.

Keeping track of the flow of genetic information can be pretty tricky. To help illustrate it, let's take a look at a highly simplified model that may be familiar to fans of role-playing games.

Chromosomes: D&D edition

Human DNA is organized into 23 segments, called **chromosomes,** and each person has two versions of each chromosome—one from their mother and one from their father.

In our simplified version of DNA, instead of 23 chromosomes, there will be just seven. In humans, each chromosome contains a huge amount of genetic code, but in our model each chromosome will control only one thing.

We'll use a version of of D&D's "d20" system of character stats in which each piece of DNA contains seven chromosomes:

1. STR
2. CON
3. DEX
4. CHR
5. WIS
6. INT
7. SEX

Six of these are the classic character stats from role-playing games: strength, constitution, dexterity, charisma, wisdom, and intelligence. The last one is the sex-determining chromosome.

Here's an example DNA "strand":

1.	STR	15
2.	CON	2
3.	DEX	1X
4.	CHR	12
5.	WIS	0.5X
6.	INT	14
7.	SEX	X

In our model, each chromosome contains one piece of information. This piece of information is either a stat (a number, usually between 1 and 18) or a multiplier. The last one, SEX, is the sex-determining chromosome, which, as with real human genetics, can be "X" or "Y."

Just like in real life, each person has two sets of chromosomes—one from their mother and one from their father. Imagine that your genes looked like this:

		Mom's DNA	Dad's DNA
1.	STR	15	5
2.	CON	2X	12
3.	DEX	1X	14
4.	CHR	12	1.5X
5.	WIS	0.5X	16
6.	INT	14	15
7.	SEX	X	X

The combination of these two sets of stats determines a person's characteristics. Here's the simple rule for combining stats in our system:

If you have a **number for both versions of a chromosome,** you get the bigger number as your stat. If you have a **number on one chromosome and a multiplier on the other,** your stat is the number times the multiplier. If you have a **multiplier on both sides,** you get a stat of 1.[1]

---

1 Because 1 is the multiplicative identity.

Here's how our hypothetical character from earlier would turn out:

		Mom's DNA	Dad's DNA	Final Set
1.	STR	15	5	15
2.	CON	2X	12	24
3.	DEX	13	14	14
4.	CHR	12	1.5X	18
5.	WIS	0.5X	14	7
6.	INT	14	15	15
7.	SEX	X	X	FEMALE

When one parent contributes a multiplier and the other contributes a number, the result can be very good! This character's constitution is a superhuman 24. In fact, other than a low score in wisdom, this character has great stats all around.

Now, let's say this character (call her "Alice") meets someone else ("Bob"): Bob also has stellar stats:

Bob		Mom's DNA	Dad's DNA	Final Stat
1.	STR	13	7	13
2.	CON	5	18	18
3.	DEX	15	11	15
4	CHR	10	2X	20
5.	WIS	16	14	16
6.	INT	2X	8	16
7.	SEX	X	Y	MALE

If they have a child, each one will contribute a strand of DNA. But the strand they contribute will be a random mix of their mother and father strands. Every sperm cell—and every egg cell—contains a random combination of chromosomes from each strand. So let's say Bob and Alice make the following sperm and egg:

	Alice	Mom's DNA	Dad's DNA	Bob	Mom's DNA	Dad's DNA
1.	STR	(15)	5	STR	13	(7)
2.	CON	(2X)	12	CON	(5)	18
3.	DEX	13	(14)	DEX	15	(11)
4.	CHR	12	(1.5X)	CHR	(10)	2X
5.	WIS	0.5X	(14)	WIS	(16)	14
6.	INT	(14)	15	INT	(2X)	8
7.	SEX	(X)	X	SEX	(X)	Y

	Egg (from Alice)		Sperm (from Bob)	
1.	STR	15	STR	7
2.	CON	2X	CON	5
3.	DEX	14	DEX	11
4.	CHR	1.5X	CHR	10
5.	WIS	14	WIS	16
6.	INT	14	INT	2X
7.	SEX	X	SEX	X

If these sperm and egg combine, the child's stats will look like this:

		Egg	Sperm	Child stats
1.	STR	15	7	15
2.	CON	2X	5	10
3.	DEX	14	11	14
4.	CHR	1.5X	10	15
5.	WIS	14	16	16
6.	INT	14	2X	28
7.	SEX	X	X	FEMALE

Alice has her mother's strength and her father's wisdom. She also has superhuman intelligence, thanks to the very good 14 contributed by Alice and the multiplier contributed by Bob. Her constitution, on the other hand, is much weaker than either of her parents, since her mother's 2x multiplier could only do so much with the "5" contributed by her father.

Alice and Bob *both* had a multiplier on their paternal "charisma" chromosome. Since two multipliers together result in a stat of 1, if Alice and Bob had both contributed their multiplier, the child would have a rock-bottom CHR. Fortunately, the odds of this happening were only 1 in 4.

If the child had multipliers on both strands, the stat would have been reduced to 1. Fortunately, since multipliers are relatively rare, the odds of them lining up in two random people are low.

Now let's look at what would happen if Alice had a child with herself.

First, she'd produce a pair of sex cells, which would run the random selection process twice:

Alice's egg	Mom's DNA	Dad's DNA	Alice's sperm	Mom's DNA	Dad's DNA
1. STR	(15)	5	STR	15	(5)
2. CON	(2X)	12	CON	(2X)	12
3. DEX	13	(14)	DEX	13	(14)
4. CHR	12	(1.5X)	CHR	(12)	1.5X
5. WIS	0.5X	(14)	WIS	(0.5X)	14
6. INT	(14)	15	INT	(14)	15
7. SEX	(X)	X	SEX	X	(X)

Then the selected strands would be contributed to the child:

Alice II	Egg	Sperm	Child stats
1. STR	15	5	15
2. CON	2X	2X	1
3. DEX	14	14	14
4. CHR	1.5X	12	16
5. WIS	0.5X	14	7
6. INT	14	14	15
7. SEX	X	X	X

The child is guaranteed to be female, since there's nobody to contribute a Y chromosome.

The child also has a problem: For three of her seven stats—INT, DEX, and CON—she inherited the same chromosome on both sides. This isn't a problem for

DEX and CON, since Alice had a high score in those two categories, but in CON, she inherited a multiplier from both sides, giving her a constitution score of 1.

If someone produces a child on their own, it dramatically increases the likelihood that the child will inherit the same chromosome on both sides, and thus a double multiplier. The odds of Alice's child having a double multiplier are 58 percent—compared to the 25 percent chance for a child with Bob.

In general, if you have a child with yourself, 50 percent of your chromosomes will have the same stat on both sides. If that stat is a 1—or if it's a multiplier—the child will be in trouble, even though you might not have been. This condition, having the same genetic code on both copies of a chromosome, is called *homozygosity*.

Humans

In humans, probably the most common genetic disorder caused by inbreeding is spinal muscular atrophy (SMA). SMA causes the death of the cells in the spinal cord, and is often fatal or severely disabling.

SMA is caused by an abnormal version of a gene on chromosome 5. About 1 in 50 people have this abnormality, which means 1 in 100 people will contribute it to their children . . . and, therefore, 1 in 10,000 people (100 times 100) will inherit the defective gene from *both* parents.[2]

If a parent has a child with his- or herself, on the other hand, the chance of SMA is 1 in 400—since if he or she has a copy of the defective gene (1 in 100), there's a 1 in 4 chance it will be the child's *only* copy.

One in 400 may not sound so bad, but SMA is only the start.

DNA is complicated

DNA is source code for the most complex machine in the known universe. Each chromosome contains a staggering amount of information, and the interaction between DNA and the cell machinery around it is incredibly complicated, with countless moving parts and Mousetrap-style feedback loops. Even calling DNA "source code" sells it short—compared to DNA, our most complex programming projects are like pocket calculators.

In humans, each chromosome affects many things through a variety of mutations and variations. Some of these mutations, like the one responsible for SMA,

---

2 Some forms of SMA are actually caused by a defect in *two* genes, so in practice the statistical picture is a little more complicated.

seem to be entirely negative; the mutation responsible has no benefit. In our D&D system, it's like a chromosome having an STR of 1. If your other chromosome is normal, you'll have a normal character stat; you'll be a silent "carrier."

Other mutations, like the sickle-cell gene on chromosome 11, can provide a mix of benefit and harm. People who have the sickle-cell gene on both their copies of the chromosome suffer from **sickle-cell anemia.** However, if they have the gene on just *one* of their chromosomes, they get a surprise benefit: extra resistance to malaria.

In the D&D system, this is like a "2x" multiplier. One copy of the gene can make you stronger, but two copies—double multipliers—lead to a serious disorder.

These two diseases illustrate one reason that genetic diversity is important. Mutations pop up all over the place, but our redundant chromosomes help blunt this effect. By avoiding inbreeding, a population reduces the odds that rare and harmful mutations will pop up at the same place on both sides of the chromosome.

Inbreeding coefficient

Biologists use a number called the "inbreeding coefficient" to quantify the percentage of someone's chromosomes that are likely to be identical. A child from unrelated parents has an inbreeding coefficient of 0, while one who has a completely duplicated set of chromosomes has an inbreeding coefficient of 1.

This brings us to the answer to the original question. A child from a parent who self-fertilized would be like a clone of the parent with severe genetic damage. The parent would have all the genes the child would, but the child wouldn't have all the genes of the parent. Half the child's chromosomes would have their "partner" chromosomes replaced by a copy of themselves.

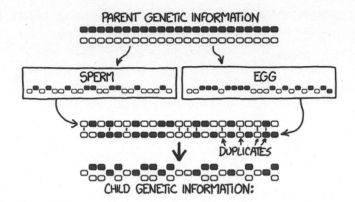

PARENT GENETIC INFORMATION

SPERM   EGG

DUPLICATES

CHILD GENETIC INFORMATION:

This means the child would have an inbreeding coefficient of 0.50. This is very high; it's what you would expect in a child of three generations of consecutive sibling marriages. According to D. S. Falconer's *Introduction to Quantitative Genetics,* an inbreeding coefficient of 0.50 would result in an average of a 22-point reduction in IQ and a 4-inch reduction in height at age ten. There would be a very good chance that the resulting fetus would not survive to birth.

This kind of inbreeding was famously exhibited by royal families attempting to keep their bloodlines "pure." The European House of Hapsburg, a family of European rulers from the mid-second millennium, was marked by frequent cousin marriages, culminating in the birth of King Charles II of Spain.

Charles had an inbreeding coefficient of 0.254, making him slightly more inbred than a child of two siblings (0.250). He suffered from extensive physical and emotional disabilities, and was a strange (and largely ineffective) king. In one incident, he reportedly ordered that the corpses of his relatives be dug up so he could look at them. His inability to bear children marked the end of that royal bloodline.

Self-fertilization is a risky strategy, which is why sex is so popular among large and complex organisms.[3] There are occasionally complex animals that reproduce asexually,[4] but this behavior is relatively rare. It typically appears in environments

---

3 Well, one of the reasons.

4 "Tremblay's Salamander" is a hybrid species of salamander that reproduces exclusively by self-fertilizing. These salamanders are an all-female species, and—strangely—have three genomes instead of two. To breed, they go through a courtship ritual with male salamanders of related species, then lay self-fertilized eggs. The male salamander gets nothing out of it; he's simply used to stimulate egg-laying.

where it's difficult to reproduce sexually, whether due to resource scarcity, population isolation . . .

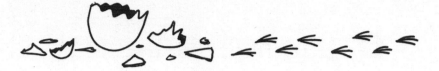

Life finds a way.

. . . or overconfident theme park operators.

HIGH THROW

Q. How high can a human throw something?

—**Irish Dave on the Isle of Man**

A. **HUMANS ARE GOOD AT** throwing things. In fact, we're great at it; no other animal can throw stuff like we can.

It's true that chimpanzees hurl feces (and, on rare occasions, stones), but they're not nearly as accurate or precise as humans. Antlions throw sand, but they don't aim it. Archerfish hunt insects by throwing water droplets, but they use specialized mouths instead of arms. Horned lizards shoot jets of blood from their eyes for distances of up to 5 feet. I don't know *why* they do this because whenever I reach the phrase "shoot jets of blood from their eyes" in an article I just stop there and stare at it until I need to lie down.

So while there are other animals that use projectiles, we're just about the only animal that can grab a random object and reliably nail a target. In fact, we're so good at it that some researchers have suggested that rock-throwing played a central role in the evolution of the modern human brain.

Throwing is hard.[1] In order to deliver a baseball to a batter, a pitcher has to release the ball at exactly the right point in the throw. A timing error of half a millisecond in either direction is enough to cause the ball to miss the strike zone.

To put that in perspective, it takes about *five* milliseconds for the fastest nerve impulse to travel the length of the arm. That means that when your arm is still rotating toward the correct position, the signal to release the ball is already at your wrist. In terms of timing, this is like a drummer dropping a drumstick from the tenth story and hitting a drum on the ground *on the correct beat.*

We seem to be much better at throwing things forward than throwing them upward.[2] Since we're going for maximum height, we could use projectiles that curve upward when you throw them forward; the Aerobie Orbiters I had when I was a kid often got stuck in the highest treetops.[3] But we could also sidestep the whole problem by using a device like this one:

A mechanism for hitting yourself in the head with a baseball after a four-second delay

We could use a springboard, a greased chute, or even a dangling sling—anything that redirects the object upward without adding to or subtracting from its speed. Of course, we could also try this:

---

1 Citation: my Little League career.
2 Counterexample: my Little League career.
3 Where they remained forever.

I ran through the basic aerodynamic calculations for a baseball thrown at various speeds. I will give these heights in units of giraffes:

The average person can probably throw a baseball at least three giraffes high:

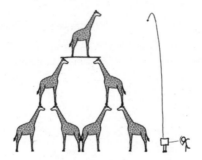

Someone with a reasonably good arm could manage five:

A pitcher with an 80 mph fastball could manage ten giraffes:

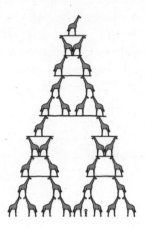

Aroldis Chapman, the holder of the world record for fastest recorded pitch (105 mph), could in theory launch a baseball 14 giraffes high:

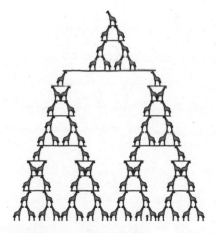

But what about projectiles other than a baseball? Obviously, with the aid of tools like slings, crossbows, or the curved *xistera* scoops in jai alai, we can launch projectiles much faster than that. But for this question, let's assume we stick to bare-handed throwing.

A baseball is probably not the ideal projectile, but it's hard to find speed data on other kinds of thrown objects. Fortunately, a British javelin thrower named Roald Bradstock held a "random object throwing competition," in which he threw everything from dead fish to an actual kitchen sink. Bradstock's experience gives us a lot of useful data.[4] In particular, it suggests a potentially superior projectile: a golf ball.

Few professional athletes have been recorded throwing golf balls. Fortunately, Bradstock has, and he claims a record throw of 170 yards. This involved a running start, but even so, it's reason to think that a golf ball might work better than a baseball. From a physics standpoint, it makes sense; the limiting factor in baseball pitches is the torque on the elbow, and the lighter golf ball might allow the pitching arm to move slightly faster.

The speed improvement from using a golf ball instead of a baseball would probably not be very large, but it seems plausible that a professional pitcher with some time to practice could throw a golf ball faster than a baseball.

If so, based on aerodynamic calculations, Aroldis Chapman could probably throw a golf ball about sixteen giraffes high:

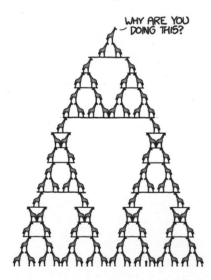

---

4 And a lot of other data, too.

This is probably about the maximum possible altitude for a thrown object.

   . . . unless you count the technique by which any five-year-old can beat all these records easily.

Q. How close would you have to be to a supernova to get a lethal dose of neutrino radiation?

—Dr. Donald Spector

A. **THE PHRASE "LETHAL DOSE** of neutrino radiation" is a weird one. I had to turn it over in my head a few times after I heard it.

If you're not a physics person, it might not sound odd to you, so here's a little context for why it's such a surprising idea:

Neutrinos are ghostly particles that barely interact with the world at all. Look at your hand—there are about a trillion neutrinos from the Sun passing through it every second.

Okay, you can stop looking at your hand now.

The reason you don't notice the neutrino flood is that neutrinos mostly ignore ordinary matter. On average, out of that massive flood, only one neutrino will "hit" an atom in your body every few years.[1]

In fact, neutrinos are so shadowy that the entire Earth is transparent to them; nearly all of the Sun's neutrino steam goes straight through it unaffected. To detect neutrinos, people build giant tanks filled with hundreds of tons of target material in the hopes that they'll register the impact of a single solar neutrino.

This means that when a particle accelerator (which produces neutrinos) wants to send a neutrino beam to a detector somewhere else in the world, all it has to do is point the beam at the detector—even if it's on the other side of the Earth!

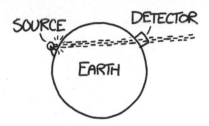

That's why the phrase "lethal dose of neutrino radiation" sounds weird—it mixes scales in an incongruous way. It's like the idiom "knock me over with a feather" or the phrase "football stadium filled to the brim with ants."[2] If you have a math background, it's sort of like seeing the expression "$\ln(x)^e$"—it's not that, taken literally, it doesn't make sense—it's that you can't imagine a situation where it would apply.[3]

Similarly, it's hard to produce enough neutrinos to get even a single *one* of them to interact with matter; it's strange to imagine a scenario in which there'd be enough of them to hurt you.

Supernovae provide that scenario.[4] Dr. Spector, the Hobart and William Smith Colleges physicist who asked me this question, told me his rule of thumb for estimating supernova-related numbers: However big you think supernovae are, they're bigger than that.

---

1 Less often if you're a child, since you have fewer atoms to be hit. Statistically, your first neutrino interaction probably happens somewhere around age ten.

2 Which would still be less than 1 percent of the ants in the world.

3 If you want to be mean to first-year calculus students, you can ask them to take the derivative of $\ln(x)^e$ dx. It looks like it should be "1" or something, but it's not.

4 "Supernovas" is also fine. "Supernovii" is discouraged.

Here's a question to give you a sense of scale. Which of the following would be brighter, in terms of the amount of energy delivered to your retina:

A supernova, seen from as far away as the Sun is from the Earth, or the detonation of a hydrogen bomb *pressed against your eyeball*?

Can you hurry up and set it off? This is heavy.

Applying Dr. Spector's rule of thumb suggests that the supernova is brighter. And indeed, it is . . . by *nine orders of magnitude*.

That's why this is a neat question—supernovae are unimaginably huge and neutrinos are unimaginably insubstantial. At what point do these two unimaginable things cancel out to produce an effect on a human scale?

A paper by radiation expert Andrew Karam provides an answer. It explains that during certain supernovae, the collapse of a stellar core into a neutron star, 10^{57} neutrinos can be released (one for every proton in the star that collapses to become a neutron).

Karam calculates that the neutrino radiation dose at a distance of 1 parsec[5] would be around half a nanosievert, or 1/500th the dose from eating a banana.[6]

A fatal radiation dose is about 4 sieverts. Using the inverse-square law, we can calculate the radiation dose:

$$0.5 \text{ nanosieverts} \times \left(\frac{1 \text{ parsec}}{x}\right)^2 = 5 \text{ sieverts}$$

$$x = 0.00001118 \text{ parsecs} = 2.3 \text{ AU}$$

That's a little more than the distance between the Sun and Mars.

Core-collapse supernovae happen to giant stars, so if you observed a super-

---

5 3.262 light-years, or a little less than the distance from here to Alpha Centauri.
6 "Radiation Dose Chart," *http://xkcd.com/radiation.*

nova from that distance, you'd probably be inside the outer layers of the star that created it.

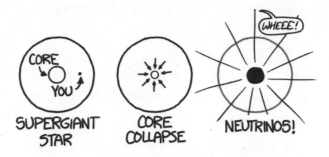

*GRB 080319B was the most violent event ever observed—
especially for the people who were floating right next to it with surfboards.*

The idea of neutrino radiation damage reinforces just how big supernovae are. If you observed a supernova from 1 AU away—and you somehow avoided being incinerated, vaporized, and converted to some type of exotic plasma—even the flood of ghostly neutrinos would be dense enough to kill you.

If it's going fast enough, a feather can *absolutely* knock you over.

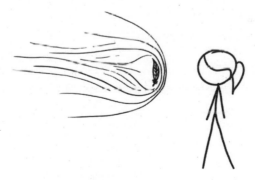

WEIRD (AND WORRYING) QUESTIONS
FROM THE WHAT IF? INBOX, #8

Q. A toxin blocks the ability of the nephron tubule reabsorption but does not affect filtration. What are the possible short-term effects of this toxin?

—Mary

Q. If a Venus fly trap could eat a person, about how long would it take for the human to be fully de-juiced and absorbed?

—Jonathan Wang

Q. How fast can you hit a speed bump while driving and live?

—**Myrlin Barber**

A. SURPRISINGLY FAST.

First, a disclaimer. After reading this answer, don't try to drive over speed bumps at high speeds. Here are some reasons:

- You could hit and kill someone.
- It can destroy your tires, suspension, and potentially your entire car.
- Have you *read* any of the other answers in this book?

If that's not enough, here are some quotes from medical journals on spinal injury from speed bumps.

Examination of the thoracolumbar X-ray and computed tomography displayed compression fractures in four patients . . . Posterior instrumentation was applied . . . All patients recovered well except for the one with cervical fracture.

L1 was the most frequently fractured vertebra (23/52, 44.2 percent).

Incorporation of the buttocks with realistic properties diminished the first vertical natural frequency from ~12 to 5.5 Hz, in agreement with the literature.

(That last one isn't directly related to speed bump injuries, but I wanted to include it anyway.)

Regular little speed bumps probably won't kill you

Speed bumps are designed to make drivers slow down. Going over a typical speed bump at 5 miles per hour results in a gentle bounce,[1] while hitting one at 20 delivers a sizable jolt. It's natural to assume that hitting a speed bump at 60 would deliver a proportionally larger jolt, but it probably wouldn't.

As those medical quotes attest, it's true that people are occasionally injured by speed bumps. However, nearly all of those injuries happen to a very specific category of people: those sitting in hard seats in the backs of buses, riding on poorly maintained roads.

When you're driving a car, the two main things protecting you from bumps in the road are the tires and the suspension. No matter how fast you hit a speed bump, unless the bump is large enough to hit the frame of the car, enough of the jolt will be absorbed by these two systems that you probably won't be hurt.

Absorbing the shock won't necessarily be *good* for those systems. In the case of the tires, they may absorb it by exploding.[2] If the bump is large enough to hit the wheel rims, it may permanently damage a lot of important parts of the car.

The typical speed bump is between 3 and 4 inches tall. That's also about how thick an average tire's cushion is (the separation between the bottom of the rims and the ground).[3] This means that if a car hits a small speed bump, the rim won't actually touch the bump; the tire will just be compressed.

The typical sedan has a top speed of around 120 miles per hour. Hitting a speed bump at that speed would, in one way or another, probably result in losing control of the car and crashing.[4] However, the jolt *itself* probably wouldn't be fatal.

---

1 Like anyone with a physics background, I do all my calculations in SI units, but I've gotten too many US speeding tickets to write this answer in anything but miles per hour; it's just been burned into my brain. Sorry!

2 Just Google "hit a curb at 60."

3 There are cars everywhere. Go outside with a ruler and check.

4 At high speeds, you can easily lose control even without hitting a bump. Joey Huneycutt's 220-mph crash left his Camaro a burned-out hulk.

If you hit a larger speed bump—like a speed hump or speed table—your car might not fare so well.

How fast would you have to go to definitely die?

Let's consider what would happen if a car were going *faster* than its top speed. The average modern car is limited to a top speed of around 120 mph, and the fastest can go about 200.

While most passenger cars have some kind of artificial speed limits imposed by the engine computer, the ultimate physical limit to a car's top speed comes from air resistance. This type of drag increases with the square of speed; at some point, a car doesn't have enough engine power to push through the air any faster.

If you *did* force a sedan to go faster than its top speed—perhaps by reusing the magical accelerator from the relativistic baseball—the speed bump would be the least of your problems.

Cars generate lift. The air flowing around a car exerts all kinds of forces on it.

Where did all these arrows come from?

The lift forces are relatively minor at normal highway speeds, but at higher speeds they become substantial.

In a Formula One car equipped with airfoils, this force pushes downward, holding the car against the track. In a sedan, they lift it up.

Among NASCAR fans, there's frequently talk of a 200-mph "liftoff speed" if the car starts to spin. Other branches of auto racing have seen spectacular backflip crashes when the aerodynamics don't work out as planned.

The bottom line is that in the range of 150–300 mph, a typical sedan would lift off the ground, tumble, and crash . . . before you even hit the bump.

BREAKING: Child, Unidentified Creature in Bicycle Basket Hit and Killed by Car

If you kept the car from taking off, the force of the wind at those speeds would strip away the hood, side panels, and windows. At higher speeds, the car itself would be disassembled, and might even burn up like a spacecraft reentering the atmosphere.

What's the ultimate limit?

In the state of Pennsylvania, drivers may see \$2 added to their speeding ticket for every mile per hour by which they break the speed limit.

Therefore, if you drove a car over a Philadelphia speed bump at 90 percent of the speed of light, in addition to destroying the city . . .

. . . you could expect a speeding ticket of \$1.14 billion.

Q. If two immortal people were placed on opposite sides of an uninhabited Earthlike planet, how long would it take them to find each other?
100,000 years?
1,000,000 years?
100,000,000,000 years?
—Ethan Lake

- -

A. **WE'LL START WITH THE** simple, physicist-style[1] answer: 3000 years.

That's about how long it would take two people to find each other, assuming that they were walking around at random over a sphere for 12 hours per day and had to get within a kilometer to see each other.

_____

1 Assuming a spherical immortal human in a vacuum . . .

We can immediately see some problems with this model.[2] The simplest problem is the assumption that you can always see someone if they come within a kilometer of you. That's possible under only the most ideal circumstances; a person walking along a ridge might be visible from a kilometer away, whereas in a thick forest during a rainstorm, two people could pass within a few meters without seeing each other.

We could try to calculate the average visibility across all parts of the Earth, but then we run into another question: Why would two people who are trying to find each other spend time in a thick jungle? It would seem to make more sense for both of them to stay in flat, open areas where they could easily see and be seen.[3]

Once we start considering the psychology of our two people, our spherical-immortal-in-a-vacuum model is in trouble.[4] Why should we assume our people will walk around randomly at all? The optimal strategy might be something totally different.

What strategy *would* make the most sense for our lost immortals?

---

2 Like, what happened to all the other people? Are they okay?

3 Although the visibility calculation *does* sounds fun. I know what I'm doing next Saturday night!

4 Which is why we usually try not to consider things like that.

If they have time to plan beforehand, it's easy. They can arrange to meet at the North or South Pole, or—if those turn out to be unreachable—at the highest point on land, or the mouth of the longest river. If there's any ambiguity, they can just travel between all the options at random. They have plenty of time.

If they don't have a chance to communicate beforehand, things get a little harder. Without knowing the other person's strategy, how do you know what *your* strategy should be?

There's an old puzzle, from before the days of cell phones, that goes something like this:

> *Suppose you're meeting a friend in an American town that neither of you have been to before. You don't have a chance to plan a meeting place beforehand. Where do you go?*

The author of the puzzle suggested that the logical solution would be to go to the town's main post office and wait at the main receiving window, where out-of-town packages arrive. His logic was that it's the only place that every town in the US has exactly one of, and which everyone would know where to find.

To me, that argument seems a little weak. More importantly, it doesn't hold up experimentally. I've asked that question to a number of people, and none of them suggested the post office. The original author of that puzzle would be waiting in the mailroom alone.

AT LEAST I HAVE
ALL THE MAIL I
COULD EVER EAT.

Our lost immortals have it tougher, since they don't know anything about the geography of the planet they're on.

Following the coastlines seems like a sensible move. Most people live near water, and it's much faster to search along a line than over a plane. If your guess turns out to be wrong, you won't have wasted much time compared to having searched the interior first.

Walking around the average continent would take about five years, based on typical width-to-coastline-length ratios for Earth land masses.[5]

Let's assume you and the other person are on the same continent. If you

---

5 Of course, some areas would present a challenge. Louisiana's bayous, the Caribbean's mangrove forests, and Norway's fjords would all make for slower walking than a typical beach.

both walk counterclockwise, you could circle forever without finding each other. That's no good.

A different approach would be to make a complete circle counterclockwise, then flip a coin. If it comes up heads, circle counterclockwise again. If tails, go clockwise. If you're both following the same algorithm, this would give you a high probability of meeting within a few circuits.

The assumption that you're both using the same algorithm is probably optimistic. Fortunately, there's a better solution: Be an ant.

Here's the algorithm that I would follow (if you're ever lost on a planet with me, keep this in mind!):

If you have no information, walk at random, leaving a trail of stone markers, each one pointing to the next. For every day that you walk, rest for three. Periodically mark the date alongside the cairn. It doesn't matter how you do this, as long as it's consistent. You could chisel the number of days into a rock, or lay out rocks to plot the number.

If you come across a trail that's newer than any you've seen before, start following it as fast as you can. If you lose the trail and can't recover it, resume leaving your own trail.

You don't have to come across the other player's current location; you simply have to come across a location where they've been. You can still chase one another in circles, but as long as you move more quickly when you're following a trail than when you're leaving one, you'll find each other in a matter of years or decades.

And if your partner isn't cooperating—perhaps they're just sitting where they started and waiting for you—then you'll get to see some neat stuff.

ORBITAL SPEED

Q. What if a spacecraft slowed down on reentry to just a few miles per hour using rocket boosters like the Mars sky crane? Would it negate the need for a heat shield?

—Brian

Q. Is it possible for a spacecraft to control its reentry in such a way that it avoids the atmospheric compression and thus would not require the expensive (and relatively fragile) heat shield on the outside?

—Christopher Mallow

Q. Could a (small) rocket (with payload) be lifted to a high point in the atmosphere where it would only need a small rocket to get to escape velocity?

—Kenny Van de Maele

A. THE ANSWERS TO THESE questions all hinge on the same idea. It's an idea I've touched on in other answers, but right now I want to focus on it specifically:

The reason it's hard to get to orbit **isn't** that space is high up.

It's hard to get to orbit because you have to go so *fast*.

Space isn't like this:

Not actual size.

Space is like *this:*

You know what, sure, actual size.

Space is about 100 kilometers away. That's far away—I wouldn't want to climb a ladder to get there—but it isn't *that* far away. If you're in Sacramento, Seattle, Canberra, Kolkata, Hyderabad, Phnom Penh, Cairo, Beijing, central Japan, central Sri Lanka, or Portland, space is closer than the sea.

Getting to space is easy.[1] It's not, like, something you could do in your car, but it's not a huge challenge. You could get a person to space with a rocket the size of a telephone pole. The X-15 aircraft reached space just by going fast and then steering up.[2,3]

You will go to space today, and then you will quickly come back.

But *getting* to space is easy. The problem is *staying* there.

Gravity in low Earth orbit is almost as strong as gravity on the surface. The Space Station hasn't escaped Earth's gravity at all; it's experiencing about 90 percent the pull that we feel on the surface.

To avoid falling back into the atmosphere, you have to go sideways **really, really fast.**

The speed you need to stay in orbit is about 8 kilometers per second.[4] Only a fraction of a rocket's energy is used to lift up out of the atmosphere; the vast majority of it is used to gain orbital (sideways) speed.

This leads us to the central problem of getting into orbit: **Reaching orbital speed takes much more fuel than reaching orbital height.** Getting a ship up to 8 km/s takes a *lot* of booster rockets. Reaching orbital speed is hard enough; reaching orbital speed while carrying enough fuel to slow back down would be completely impractical.[5]

These outrageous fuel requirements are why every spacecraft entering an atmosphere has braked using a heat shield instead of rockets—slamming into the

---

1 Specifically, low Earth orbit, which is where the International Space Station is and where shuttles could go.
2 The X-15 reached 100 km on two occasions, both when flown by Joe Walker.
3 Make sure to remember to steer up and not down, or you will have a bad time.
4 It's a little less if you're in the higher region of low Earth orbit.
5 This exponential increase is the central problem of rocketry: The fuel required to increase your speed by 1 km/s multiplies your weight by about 1.4. To get into orbit, you need to increase your speed to 8 km/s, which means you'll need a lot of fuel: $1.4 \times 1.4 \times 1.4 \times 1.4 \times 1.4 \times 1.4 \times 1.4 \times 1.4 \approx 15$ times the original weight of your ship. Using a rocket to slow down carries the same problem: Every 1 km/s decrease in speed multiplies your starting mass by that same factor of 1.4. If you want to slow all the way down to zero—and drop gently into the atmosphere—the fuel requirements multiply your weight by 15 again.

air is the most practical way to slow down. (And to answer Brian's question, the Curiosity rover was no exception to this; although it used small rockets to hover when it was near the surface, it first used air-braking to shed the majority of its speed.)

How fast is 8 km/s, anyway?

I think the reason for a lot of confusion about these issues is that when astronauts are in orbit, it doesn't seem like they're moving that fast; they look like they're drifting slowly over a blue marble.

But 8 km/s is *blisteringly* fast. When you look at the sky near sunset, you can sometimes see the ISS go past . . . and then, 90 minutes later, see it go past again.[6] In those 90 minutes, it's circled the entire world.

The ISS moves so quickly that if you fired a rifle bullet from one end of a football field,[7] the International Space Station could cross the length of the field before the bullet traveled 10 yards.[8]

Let's imagine what it would look like if you were speed-walking across the Earth's surface at 8 km/s.

To get a better sense of the pace at which you're traveling, let's use the beat of a song to mark the passage of time.[9] Suppose you started playing the 1988 song by The Proclaimers, "I'm Gonna Be (500 Miles)." That song is about 131.9 beats per minute, so imagine that with every beat of the song, you move forward more than 2 miles.

In the time it took to sing the first line of the chorus, you could walk from the Statue of Liberty all the way to the Bronx.

You'd be moving at about 15 subway stops per second.

It would take you about two lines of the chorus (16 beats of the song) to cross the English Channel between London and France.

The song's length leads to an odd coinci-

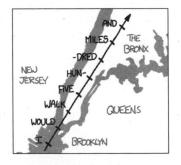

---

6 There are some good apps and online tools to help you spot the station, along with other neat satellites.
7 Either kind.
8 This type of play is legal in Australian rules football.
9 Using song beats to help measure the passage of time is a technique also used in CPR training, where the song "Stayin' Alive" is used.

dence. The interval between the start and the end of "I'm Gonna Be" is 3 minutes and 30 seconds, and the ISS is moving at 7.66 km/s.

This means that if an astronaut on the ISS listens to "I'm Gonna Be," in the time between the first beat of the song and the final lines . . .

JUST TO BE THE MAN WHO WALKED A THOUSAND MILES TO FALL DOWN AT YOUR DOOR

. . . they will have traveled just about *exactly* 1000 miles.

Q. When – if ever – will the bandwidth of the Internet surpass that of FedEx?

—Johan Öbrink

Never underestimate the bandwidth of a station wagon full of tapes hurtling down the highway.

–Andrew Tanenbaum, 1981

A. IF YOU WANT TO transfer a few hundred gigabytes of data, it's generally faster to FedEx a hard drive than to send the files over the Internet. This isn't a new idea—it's often dubbed "SneakerNet"—and it's even how Google transfers large amounts of data internally.

But will it always be faster?

Cisco estimates that total Internet traffic currently averages 167 terabits per second. FedEx has a fleet of 654 aircraft with a lift capacity of 26.5 million pounds daily. A solid-state laptop drive weighs about 78 grams and can hold up to a terabyte.

That means FedEx is capable of transferring 150 exabytes of data per day, or 14 petabits per second—almost a hundred times the current throughput of the Internet.

If you don't care about cost, this 10-kilogram shoebox can hold a lot of Internet.

TOP-END LAPTOP DRIVES: 136
STORAGE: 136 TERABYTES
COST: $130,000
(PLUS $40 FOR THE SHOES)

We can improve the data density even further by using microSD cards:

MICROSD CARDS: 25,000
STORAGE: 1.6 PETABYTES
RETAIL COST: $1.2 MILLION

Those thumbnail-sized flakes have a storage density of up to 160 terabytes per kilogram, which means a FedEx fleet loaded with microSD cards could transfer about 177 petabits per second, or 2 zettabytes per day—a thousand times the Internet's current traffic level. (The infrastructure would be interesting—Google would need to build huge warehouses to hold a massive card-processing operation.)

Cisco estimates Internet traffic is growing at about 29 percent annually. At that rate, we'll hit the FedEx point in 2040. Of course, the amount of data we can fit on a drive will have gone up by then, too. The only way to actually reach the FedEx point is if transfer rates grow much faster than storage rates. In an intuitive sense, this seems unlikely, since storage and transfer are fundamentally linked—all that data is coming from somewhere and going somewhere—but there's no way to predict usage patterns for sure.

While FedEx is big enough to keep up with the next few decades of actual usage, there's no technological reason we can't build a connection that beats them on bandwidth. There are experimental fiber clusters that can handle over a petabit per second. A cluster of 200 of those would beat FedEx.

If you recruited the entire US freight industry to move SD cards for you, the throughput would be on the order of 500 exabits—half a zettabit—per second. To match that transfer rate digitally, you'd need to take half a million of those petabit cables.

So the bottom line is that for raw bandwidth of FedEx, the Internet will probably never beat SneakerNet. However, the virtually infinite bandwidth of a FedEx-based Internet would come at the cost of 80,000,000-millisecond ping times.

FREE FALL

Q. What place on Earth would allow you to free-fall the longest by jumping off it? What about using a squirrel suit?

—Dhash Shrivathsa

A. THE LARGEST PURELY VERTICAL drop on Earth is the face of Canada's Mount Thor, which is shaped like this:

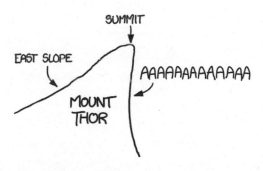

Source: AAAAAAAAAAAAAAAAAAAAAAAAAAAAAAAAAAAAAAAAAAAAAAAAAAAAAAAAAAAAAAAAAAAA
AAAAAAAAAAAAAAAAAAAAAAAAAAAAAAAAAAAAAAAAAAAAAAAAAAAAAAAAAAAAAAAAAAAA

To make this scenario a little less gruesome, let's suppose there's a pit at the bottom of the cliff filled with something fluffy—like cotton candy—to safely break your fall.

Would this work? You'll have to wait for book two . . .

A human falling with arms and legs outstretched has a terminal velocity in the neighborhood of 55 meters per second. It takes a few hundred meters to get up to speed, so it would take you a little over 26 seconds to fall the full distance.

What can you do in 26 seconds?

For starters, it's enough time to get all the way through the original Super Mario World 1-1, assuming you have perfect timing and take the shortcut through the pipe.

It's also long enough to miss a phone call. Sprint's ring cycle—the time the phone rings before going to voicemail—is 23 seconds.[1]

If someone called your phone, and it started ringing the moment you jumped, it would go to voicemail three seconds before you reached the bottom.

> I'M SORRY I MISSED YOUR CALL, BUT IF YOU STAND AT THE FOOT OF MOUNT THOR, I'LL GET BACK TO YOU *EXTREMELY* SOON.

On the other hand, if you jumped off Ireland's 210-meter Cliffs of Moher, you would be able to fall for only about eight seconds—or a little more, if the updrafts were strong. That's not very long, but according to River Tam, given adequate vacuuming systems it might be enough time to drain all the blood from your body.

So far, we've assumed you're falling vertically. But you don't have to.

Even without any special equipment, a skilled skydiver—once he or she gets

---

1 For those keeping score, that means Wagner's is 2,350 times longer.

up to full speed—can glide at almost a 45-degree angle. By gliding away from the base of the cliff, you could conceivably extend your fall substantially.

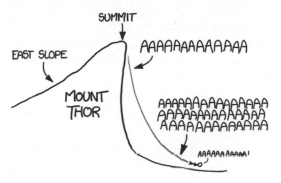

AAAAAAAAAAAAAAAAAAAAAAAAAAAAAAAAAAAAAAAAAAAAAA ::gasp::
AAAAAAAAAAAAAAAAAAAAAAAAAAAAAAAAAAAAAAAAAAAAAAAAAAAA

It's hard to say exactly how far; in addition to the local terrain, it depends heavily on your choice of clothes. As a comment on a BASE jumping records wiki puts it,

> *The record for longest [fall time] without a wingsuit is hard to find since the line between jeans and wingsuits has blurred since the introduction of more advanced . . . apparel.*

Which brings us to wingsuits—the halfway point between parachute pants and parachutes.

Wingsuits let you fall much more slowly. One wingsuit operator posted tracking data from a series of jumps. It shows that in a glide, a wingsuit can lose altitude as slowly as 18 meters per second—a huge improvement over 55.

Even ignoring horizontal travel, that would stretch out our fall to over a minute. That's long enough for a chess game. It's also long enough to sing the first verse of—appropriately enough—REM's "It's the End of the World as We Know It," followed by—less appropriately—the entire breakdown from the end of the Spice Girls' "Wannabe."

♪ SO HERE'S THE STORY
FROM A TO Z
YOU WANNA GET WITH ME
YOU GOTTA LISTEN CAREFULLY

When we include the higher cliffs opened up by horizontal glides, the times get even longer.

There are a lot of mountains that could probably support very long wingsuit flights. For example, Nanga Parbat, a mountain in Pakistan, has a drop of more than 3 kilometers at a fairly steep angle. (Surprisingly, a wingsuit still works fine in such thin air, though the jumper would need oxygen, and it would glide a little faster than normal.)

So far, the record for longest wingsuit BASE jump is held by Dean Potter, who jumped from the Eiger—a mountain in Switzerland—and flew for three minutes and twenty seconds.

What could you do with three minutes and twenty seconds?

Suppose we recruit Joey Chestnut and Takeru Kobayashi, the world's top competitive eaters.

If we can find a way for them to operate wingsuits while eating at full speed, and they jumped from the Eiger, they could—in theory—finish as many as 45 hot dogs between them before reaching the ground . . .

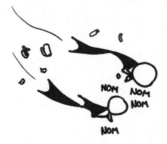

. . . which would, if nothing else, earn them what just might be the strangest world record in history.

Q. Could you survive a tidal wave by submerging yourself in an in-ground pool?

—Chris Muska

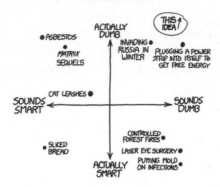

ACTUALLY DUMB

THIS IDEA

• ASBESTOS

• MATRIX SEQUELS

INVADING RUSSIA IN WINTER •

PLUGGING A POWER STRIP INTO ITSELF TO GET FREE ENERGY •

CAT LEASHES •

SOUNDS SMART ←→ SOUNDS DUMB

CONTROLLED FOREST FIRES •

LASER EYE SURGERY •

• SLICED BREAD

PUTTING MOLD ON INFECTIONS •

ACTUALLY SMART

Q. If you are in free fall and your parachute fails, but you have a Slinky with extremely convenient mass, tension, etc., would it be possible to save yourself by throwing the Slinky upward while holding on to one end of it?

—Varadarajan Srinivasan

SPARTA

Q. In the movie *300* they shoot arrows up into the sky and they seemingly blot out the sun. Is this possible, and how many arrows would it take?

—Anna Newell

A. IT'S PRETTY HARD TO make this work.

Attempt 1

Longbow archers can fire eight to ten arrows per minute. In physics terms, a longbow archer is an arrow generator with a frequency of 150 millihertz.

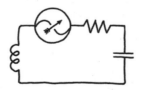

Each arrow spends only a few seconds in the air. If an arrow's average time over the battlefield is three seconds, then about 50 percent of all archers have arrows in the air at any given time.

Each arrow intercepts about 40 cm² of sunlight. Since archers have arrows in the air only half the time, each blocks an average of 20 cm² of sunlight.

If the archers are packed in rows, with two archers per meter and a row every meter and a half, and the archer battery is 20 rows (30 meters) deep, then for every meter of width . . .

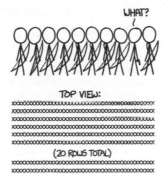

. . . there will be 18 arrows in the air.

18 arrows will block only about 0.1 percent of the Sun from the firing range. We need to improve on this.

Attempt 2

First, we can pack the archers more tightly. If they stand with the density of a mosh pit crowd,[1] we can triple the number of archers per square foot. Sure, it will make firing awkward, but I'm sure they can figure it out.

We can expand the depth of the firing column to 60 meters. That gives us a density of 130 archers per meter.

How fast can they fire?

In the extended edition of the 2001 film *Lord of the Rings: The Fellowship of*

---

1 Rule of thumb: One person per square meter is a light crowd, four people per square meter is a mosh pit.

the Ring, there's a scene where a group of orcs[2] charge at Legolas, and Legolas draws and fires arrows in rapid succession, felling the attackers with one shot each before they reach him.

The actor playing Legolas, Orlando Bloom, couldn't really fire arrows that quickly. He was actually dry-firing an empty bow; the arrows were added using CGI. Since this fire rate appeared, to the audience, to be impressively fast but not physically implausible, it provides a convenient upper limit for our calculations.

Let's assume we can train our archers to replicate Legolas's fire rate of seven arrows in eight seconds.

In that case, our column of archers (firing an impossible 339 arrows per meter) will still block out only 1.56 percent of the sunlight passing through them.

Attempt 3

Let's dispense with the bows entirely and give our archers arrow-firing Gatling bows. If they can fire 70 arrows per second, that adds up to 110 square meters of arrows per 100 square meters of battlefield! Perfect.

But there's a problem. Even though the arrows have a total cross-sectional area of 100 meters, some of them shadow each other.

The formula for the fraction of ground coverage by a large number of arrows, some of which overlap each other, is this:

$$\left(1 - \frac{\text{Arrow area}}{\text{Ground area}} \right)^{\text{Arrow count}}$$

With 110 square meters of arrows, you'll cover only two-thirds of the battle-field. Since our eyes judge brightness on a logarithmic scale, reducing the Sun's brightness to a third of its normal value will be seen as a slight dimming; certainly not "blotting it out."

With an even more unrealistic fire rate, we could make it work. If the guns release 300 arrows per second, they would block out 99 percent of the sunlight reaching the battlefield.

---

2 Strictly speaking, they were Uruk-Hai, not typical orcs. The precise nature and origin of the Uruk-Hai is a little tricky. Tolkien suggested that they were created by cross-breeding humans with orcs. However, in an earlier draft, published in *The Book of Lost Tales,* he instead suggests the Uruks had been born from the "sub-terranean heats and slimes of the Earth." Director Peter Jackson, when deciding what to show on-screen in his film adaptation, wisely went with the latter version.

But there's an easier way.

Attempt 4

We've been making the implicit assumption that the Sun is directly overhead. That's certainly what the movie shows. But perhaps the famous boast was based on a plan to attack at dawn.

If the Sun were low on the eastern horizon, and the archers were firing north, then the light could have to pass through the entire column of arrows, potentially multiplying the shadow effect a thousandfold.

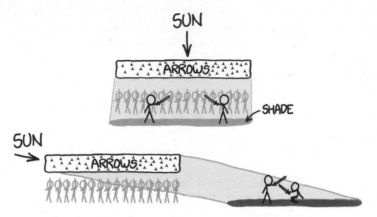

Of course, the arrows wouldn't be aimed anywhere near the enemy soldiers. But, to be fair, all they said was that their arrows would blot out the Sun. They never said anything about *hitting* anyone.

And who knows; maybe, against the right enemy, that's all they need.

Q. How quickly would the oceans drain if a circular portal 10 meters in radius leading into space were created at the bottom of Challenger Deep, the deepest spot in the ocean? How would the Earth change as the water was being drained?

—Ted M

A. I WANT TO GET one thing out of the way first:

According to my rough calculations, if an aircraft carrier sank and got stuck against the drain, the pressure would easily be enough to fold it up and suck it through. Cooool.

Just how far away is this portal? If we put it near the Earth, the ocean would just fall back down into the atmosphere. As it fell, it would heat up and turn to steam, which would condense and fall right back into the ocean as rain. The energy input into the atmosphere alone would also wreak all kinds of havoc with our climate, as would the huge clouds of high-altitude steam.

So let's put the ocean-dumping portal far away—say, on Mars. (In fact, I vote we put it directly above the Curiosity rover; that way, it will finally have incontrovertible evidence of liquid water on Mars's surface.)

What happens to the Earth?

Not much. It would actually take hundreds of thousands of years for the ocean to drain.

Even though the opening is wider than a basketball court, and the water is forced through at incredible speeds, the oceans are *huge*. When you started, the water level would drop by less than a centimeter per day.

There wouldn't even be a cool whirlpool at the surface—the opening is too small and the ocean is too deep. (It's the same reason you don't get a whirlpool in the bathtub until the water is more than halfway drained.)

But let's suppose we speed up the draining by opening more drains,[1] so the water level starts to drop more quickly.

Let's take a look at how the map would change.

Here's how it looks at the start:

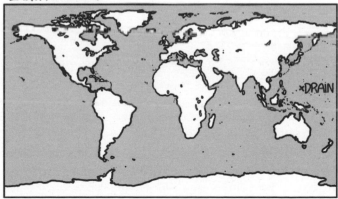

EARTH (ACTUAL SIZE):

This is a Plate Carrée projection (c.f. xkcd.com/977).

---

1 Remember to clean the whale filter every few days.

And here's the map after the oceans drop 50 meters:

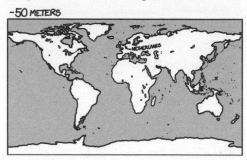

It's pretty similar, but there are a few small changes. Sri Lanka, New Guinea, Great Britain, Java, and Borneo are now connected to their neighbors.

And after 2000 years of trying to hold back the sea, the Netherlands are finally high and dry. No longer living with the constant threat of a cataclysmic flood, they're free to turn their energies toward outward expansion. They immediately spread out and claim the newly exposed land.

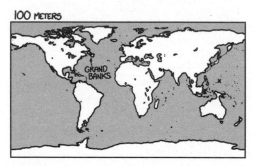

When the sea level reaches (minus) 100 meters, a huge new island off the coast of Nova Scotia is exposed—the former site of the Grand Banks.

You may start to notice something odd: Not all the seas are shrinking. The Black Sea, for example, shrinks only a little, then stops.

This is because these bodies are no longer connected to the ocean. As the water level falls, some basins cut off from the drain in the Pacific. Depending on the details of the sea floor, the flow of water out of the basin might carve a deeper channel, allowing it to continue to flow out. But most of them will eventually become landlocked and stop draining.

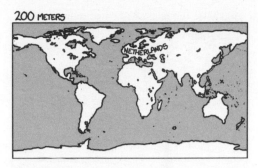

At 200 meters, the map is starting to look weird. New islands are appearing. Indonesia is a big blob. The Netherlands now control much of Europe.

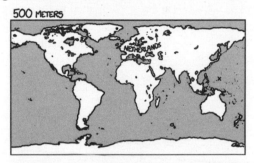

Japan is now an isthmus connecting the Korean peninsula with Russia. New Zealand gains new islands. The Netherlands expand north.

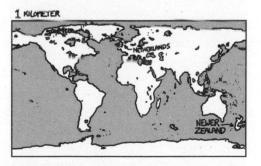

New Zealand grows dramatically. The Arctic Ocean is cut off and its water level stops falling. The Netherlands cross the new land bridge into North America.

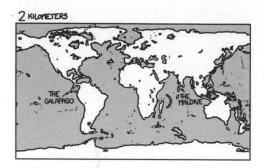

The sea has dropped by 2 kilometers. New islands are popping up left and right. The Caribbean Sea and the Gulf of Mexico are losing their connections with the Atlantic. I don't even know *what* New Zealand is doing.

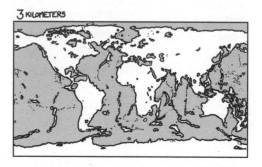

At 3 kilometers, many of the peaks of the mid-ocean ridge—the world's longest mountain range—break the surface. Vast swaths of rugged new land emerge.

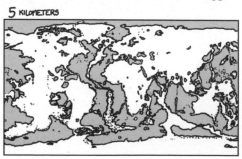

By this point, most of the major oceans have become disconnected and stopped draining. The exact locations and sizes of the various inland seas are hard to predict; this is only a rough estimate.

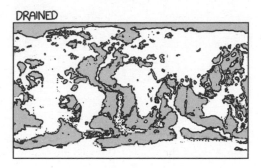

This is what the map looks like when the drain finally empties. There's a surprising amount of water left, although much of it consists of very shallow seas, with a few trenches where the water is as deep as 4 or 5 kilometers.

Vacuuming up half the oceans would massively alter the climate and ecosystems in ways that are hard to predict. At the very least, it would almost certainly involve a collapse of the biosphere and mass extinctions at every level.

But it's possible—if unlikely—that humans could manage to survive. If we did, we'd have this to look forward to:

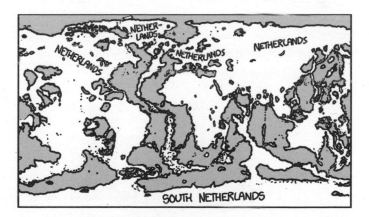

DRAIN THE OCEANS: PART II

Q. Supposing you *did* drain the oceans, and dumped the water on top of the *Curiosity* rover, how would Mars change as the water accumulated?

—Iain

- -

A. IN THE PREVIOUS ANSWER, we opened a portal at the bottom of the Mariana Trench and let the oceans drain out.

We didn't worry too much about where the oceans were draining *to*. I picked Mars; the *Curiosity* rover is working so hard to find evidence of water, so I figured we could make things easier for it.

Curiosity is sitting in Gale Crater, a round depression in the Martian surface with a peak, nicknamed Mount Sharp, in the center.

There's a lot of water on Mars. The problem is, it's frozen. Liquid water doesn't last long there, because it's too cold and there's too little air.

If you set out a cup of warm water on Mars, it'll try to boil, freeze, and sublimate, practically all at once. Water on Mars seems to want to be in *any* state except liquid.

However, we're dumping a lot of water very fast (all of it at a few degrees above 0°C), and it won't have much time to freeze, boil, or sublimate. If our portal is big enough, the water will start to turn Gale Crater into a lake, just like it would on Earth. We can use the excellent USGS Mars Topographic Map to chart the water's progress.

Here's Gale Crater at the start of our experiment:

As the flow continues, the lake fills in, burying *Curiosity* under hundreds of meters of water:

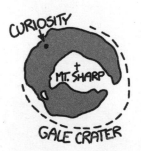

Eventually, Mount Sharp becomes an island. However, before the peak can disappear completely, the water spills over the north rim of the crater and starts flowing out across the sand.

There's evidence that—due to occasional heat waves—ice in the Martian soil occasionally melts and flows as a liquid. When this happens, the trickle of water quickly dries up before it can get very far. However, we've got a lot of ocean at our disposal.

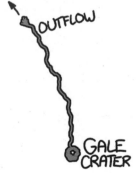

The water pools in the North Polar Basin:

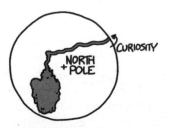

Gradually, it will fill the basin:

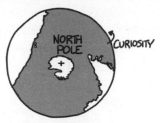

However, if we look at a map of the more equatorial regions of Mars, where the volcanoes are, we'll see that there's still a lot of land far from the water:

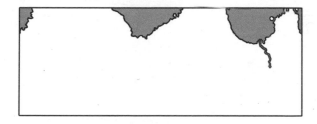

[Mercator projection; does not show the poles.]

Frankly, I think this map is kind of boring; there's not a lot going on. It's just a big empty swath of land with some ocean at the top.

Would not buy again.

We haven't come close to running out of ocean yet although there was a lot of blue on the map of the Earth at the end of our last answer, the seas that remained were shallow; most of the volume of the oceans was gone.

And Mars is much smaller than Earth, so the same volume of water will make a deeper sea.

At this point, the water fills in the Valles Marineris, creating some unusual coastlines. The map is less boring, but the terrain around the great canyons makes for some odd shapes.

The water now reaches and swallows up Spirit and Opportunity. Eventually, it breaks into the Hellas Impact Crater, the basin containing the lowest point on Mars.

In my opinion, the rest of the map is starting to look pretty good.

As the water spreads across the surface in earnest, the map splits into several large islands (and innumerable smaller ones).

The water quickly finishes covering most of the high plateaus, leaving only a few islands left.

And then, at last, the flow stops; the oceans back on Earth are drained. Let's take a closer look at the main islands:

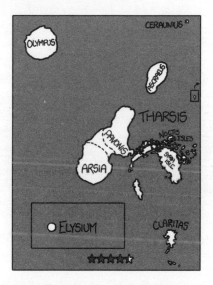

No rovers remain above water.

Olympus Mons, and a few other volcanoes, remain above water. Surprisingly, they aren't even *close* to being covered. Olympus Mons still rises well over 10 kilometers above the new sea level. Mars has some *huge* mountains.

Those crazy islands are the result of water filling in Noctis Labyrinthus (the Labyrinth of the Night), a bizarre set of canyons whose origin is still a mystery.

The oceans on Mars wouldn't last. There might be some transient greenhouse warming, but in the end, Mars is just too cold. Eventually, the oceans will freeze over, become covered with dust, and gradually migrate to the permafrost at the poles.

However, it would take a long time, and until it did, Mars would be a much more interesting place.

When you consider that there's a ready-made portal system to allow transit between the two planets, the consequences are inevitable:

TWITTER

Q. How many unique English tweets are possible? How long would it take for the population of the world to read them all out loud?

—Eric H, Hopatcong, NJ

High up in the North in the land called Svithjod, there stands a rock. It is a hundred miles high and a hundred miles wide. Once every thousand years a little bird comes to this rock to sharpen its beak. When the rock has thus been worn away, then a single day of eternity will have gone by.

—Hendrik Willem Van Loon

A. TWEETS ARE 140 CHARACTERS long. There are 26 letters in English—27 if you include spaces. Using that alphabet, there are $27^{140} \approx 10^{200}$ possible strings.

But Twitter doesn't limit you to those characters. You have all of Unicode to play with, which has room for over a million different characters. The way Twitter counts Unicode characters is complicated, but the number of possible strings could be as high as 10^{800}.

Of course, almost all of them would be meaningless jumbles of characters from a dozen different languages. Even if you're limited to the 26 English letters, the strings would be full of meaningless jumbles like "ptikobj." Eric's question was about tweets that actually say something in English. How many of those are possible?

This is a tough question. Your first impulse might be to allow only English words. Then you could further restrict it to grammatically valid sentences.

But it gets tricky. For example, "Hi, I'm Mxyztplk" is a grammatically valid sentence if your name happens to be Mxyztplk. (Come to think of it, it's just as grammatically valid if you're lying.) Clearly, it doesn't make sense to count every string that starts with "Hi, I'm . . . " as a separate sentence. To a normal English speaker, "Hi, I'm Mxyztplk" is basically indistinguishable from "Hi, I'm Mxzkqklt," and shouldn't both count. But "Hi, I'm xPoKeFaNx" is definitely recognizably different from the first two, even though "xPoKeFaNx" isn't an English word by any stretch of the imagination.

Our way of measuring distinctiveness seems to be falling apart. Fortunately, there's a better approach.

Let's imagine a language that has only two valid sentences, and every tweet must be one of the two sentences. They are:

- "There's a horse in aisle five."
- "My house is full of traps."

Twitter would look like this:

The messages are relatively long, but there's not a lot of information in each one—all they tell you is whether the person decided to send the trap message or the horse message. It's effectively a 1 or a 0. Although there are a lot of letters, for a reader who knows the pattern of the language, each tweet carries only one *bit* of information per sentence.

This example hints at a very deep idea, which is that information is fundamentally tied to the recipient's uncertainty about the message's content and his or her ability to predict it in advance.[1]

Claude Shannon—who almost singlehandedly invented modern information theory—had a clever method for measuring the information content of a language. He showed groups of people samples of typical written English that were cut off at a random point, then asked them to guess which letter came next.

It's threatening to flood our town with information!

Based on the rates of correct guesses—and rigorous mathematical analysis—Shannon determined that the information content of typical written English was 1.0 to 1.2 bits per letter. This means that a good compression algorithm should be able to compress ASCII English text—which is 8 bits per letter—to about ⅛th of its original size. Indeed, if you use a good file compressor on a .txt ebook, that's about what you'll find.

If a piece of text contains n bits of information, in a sense it means that there are 2^n different messages it can convey. There's a bit of mathematical juggling here (involving, among other things, the length of the message and something called "unicity distance"), but the bottom line is that it suggests there are on the order of about $2^{140 \times 1.1} \approx 2 \times 10^{46}$ meaningfully different English tweets, rather than 10^{200} or 10^{800}.

---

1 It also hints at a very shallow idea about there being a horse in aisle five.

Now, how long would it take the world to read them all out?

Reading 2×10^{46} tweets would take a person nearly 10^{47} seconds. It's such a staggeringly large number of tweets that it hardly matters whether it's one person reading or a billion—they won't be able to make a meaningful dent in the list in the lifetime of the Earth.

Instead, let's think back to that bird sharpening its beak on the mountaintop. Suppose that the bird scrapes off a tiny bit of rock from the mountain when it visits every thousand years, and it carries away those few dozen dust particles when it leaves. (A normal bird would probably *deposit* more beak material on the mountaintop than it would wear away, but virtually nothing else about this scenario is normal either, so we'll just go with it.)

Let's say you read tweets aloud for 16 hours a day, every day. And behind you, every thousand years, the bird arrives and scrapes off a few invisible specks of dust from the top of the hundred-mile mountain with its beak.

When the mountain is worn flat to the ground, that's the first day of eternity.

The mountain reappears and the cycle starts again for another eternal day: 365 eternal days—each one 10^{32} years long—makes an eternal year.

A hundred eternal years, in which the bird grinds away 36,500 mountains, make an eternal century.

But a century isn't enough. Nor a millennium.

Reading all the tweets takes you *ten thousand* eternal years.

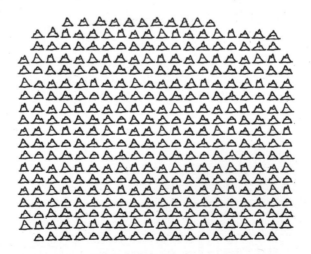

That's enough time to watch all of human history unfold, from the invention of writing to the present, with each day lasting as long as it takes for the bird to wear down a mountain.

While 140 characters may not seem like a lot, we will *never* run out of things to say.

LEGO BRIDGE

Q. How many Lego bricks would it take to build a bridge capable of carrying traffic from London to New York? Have that many Lego bricks been manufactured?

—**Jerry Petersen**

A. LET'S START WITH A less ambitious goal.

Making the connection

There have certainly been enough Lego[1] bricks to *connect* New York and London. In LEGO[2] units, New York and London are 700 million studs apart. That means that if you arranged bricks like this . . .

. . . it would take 350 million of them to connect the two cities. The bridge

---

[1] Although enthusiasts will point out it should be written "LEGO."
[2] Actually, the LEGO Group® demands that it be styled "*LEGO®*."

wouldn't be able to hold itself together or carry anything bigger than a *LEGO®*[3] minifig, but it's a start.

There have been over 400 billion Lego[4] pieces produced over the years. But how many of those are bricks that would help with a bridge, and how many are little helmet visors that get lost in the carpet?

Let's assume we're building our bridge out of the most common LeGo[5] piece—the 2x4 brick.

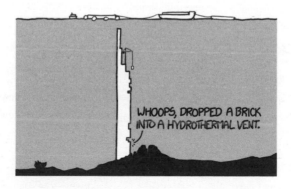

Using data provided by Dan Boger, Lego[6] kit archivist and operator of the Peeron.com Lego data site, I've come up with the following rough estimate: 1 out of every 50 to 100 pieces is a 2x4 rectangular brick. This suggests there are about 5–10 billion 2x4 bricks in existence, which is more than enough for our one-block-wide bridge.

Supporting cars

Of course, if we want to support actual traffic, we'll need to make the bridge a little wider.

We probably want to make the bridge float. The Atlantic Ocean is deep,[citation needed] and we want to avoid building 3-mile-high pylons out of Lego bricks if we can.

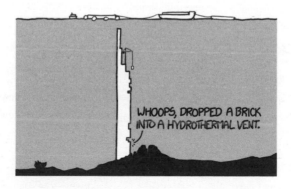

WHOOPS, DROPPED A BRICK INTO A HYDROTHERMAL VENT.

---

3 On the other hand, writers have no legal obligation to include the trademark symbol. The Wikipedia style guide mandates that it be written "Lego."

4 The Wikipedia style is not without its critics. The talk page argument over this issue featured many pages of heated arguments, including several misguided legal threats. They also debate the italics.

5 OK, *nobody* styles it this way.

6 Fine.

Lego bricks don't make a watertight seal when you connect them together,[7] and the plastic used to make them is denser than water. That's easy enough to solve; if we put a layer of sealant over the outer surface, the resulting block is substantially less dense than water.

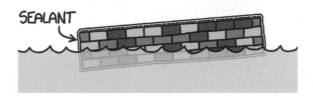

For every cubic meter of water it displaces, the bridge can carry 400 kg. A typical passenger car weighs a little under 2000 kg, so our bridge will need a minimum of 10 cubic meters of Lego supporting each passenger car.

If we make the bridge a meter thick and 5 meters wide, then it should be able to stay afloat without any trouble—although it might ride low in the water—and be sturdy enough to drive on.

Legos[8] are quite strong; according to a BBC investigation, you could stack a quarter of a million 2x2 bricks on top of each other before the bottom one collapsed.[9]

The first problem with this idea is that there aren't nearly enough Lego blocks in the world to build this kind of bridge. Our second problem is the ocean.

Extreme forces

The North Atlantic is a stormy place. While our bridge would manage to avoid the fastest-moving parts of the Gulf Stream current, it would still be subjected to powerful wind and wave forces.

How strong could we make our bridge?

Thanks to a researcher at the University of Southern Queensland named Tristan Lostroh, we have some data on the tensile strength of certain Lego joints. Their conclusion, like the BBC's, is that Lego bricks are surprisingly tough.

The optimal design would use long, thin plates overlapped with each other:

---

7 Citation: I made a Lego boat once and put it in the water and it sank :(
8 I'm going to get some angry mail about this.
9 Maybe it was a slow news day.

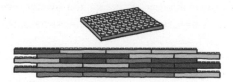

This design would be pretty strong—the tensile strength would be comparable to concrete—but not nearly strong enough. The wind, waves, and current would push the center of the bridge sideways, creating tremendous tension in the bridge.

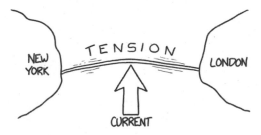

The traditional way to deal with this situation would be to anchor the bridge to the ground so it can't drift too far to one side. If we allow ourselves to use cables in addition to the Lego bricks,[10] we could conceivably tether this massive contraption to the sea floor.[11]

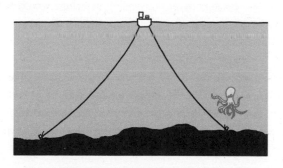

10 And sealant.
11 If we wanted to try to use Lego pieces, we could get kits that include little nylon ropes.

But the problems don't end there. A 5-meter bridge might be able to support a vehicle on a placid pond, but our bridge needs to be large enough to stay above water when waves are breaking over it. Typical wave heights on the open ocean could be several meters, so we need the deck of our bridge to be floating at least, say, 4 meters above the water.

We can make our structure more buoyant by adding air sacs and hollows, but we also need to make it wider—otherwise it will tip over. This means we have to add more anchors, with floats on those anchors to keep them from sinking. The floats create more drag, which puts more stress on the cables and pushes our structure downward, requiring more floats on the structure . . .

WAIT, THIS IS JUST THE PYLON IDEA AGAIN.

Sea floor

If we want to build our bridge down to the sea floor, we'll have a few problems. We wouldn't be able to keep the air sacs open under the pressure, so the structure would have to support its own weight. To handle the pressure from the ocean currents, we'd have to make it wider. In the end, we'd effectively be building a causeway.

As a side effect, our bridge would halt the North Atlantic Ocean circulation. According to climate scientists, this is "probably bad."[12]

Furthermore, the bridge would cross the mid-Atlantic ridge. The Atlantic sea floor is spreading outward from a seam down the middle, at a rate—in Lego

---

12 They went on to say, "Wait, *what* did you say you were trying to build?" and "How did you get in here, anyway?"

units—of one stud every 112 days. We would have to build in expansion joints, or drive out to the middle every so often and add a bunch of bricks.

Cost

Lego bricks are made of ABS plastic, which costs about a dollar per kilogram at the time of this writing. Even our simplest bridge design, the one with the kilometer-long steel tethers,[13] would cost over $5 trillion.

But consider: The total value of the London real estate market is $2.1 trillion, and transatlantic shipping rates are about $30 per ton.

This means that for less than the cost of our bridge, we could buy all the property in London and ship it, piece by piece, to New York. Then we could re-assemble it on a new island in New York Harbor, and connect the two cities with a much simpler Lego bridge.

We might even have enough left over to buy that sweet Millennium Falcon *kit.*

---

13 My favorite *Friends* episode.

Q. What is the longest possible sunset you can experience while driving, assuming we are obeying the speed limit and driving on paved roads?

—**Michael Berg**

- -

A. TO ANSWER THIS, WE have to be sure what we mean by "sunset." This is a sunset:

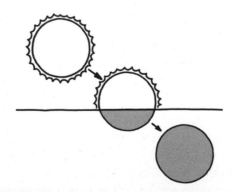

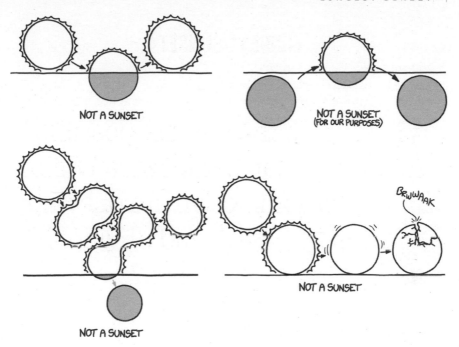

NOT A SUNSET

NOT A SUNSET
(FOR OUR PURPOSES)

BRWWAAK

NOT A SUNSET

NOT A SUNSET

Sunset starts the instant the Sun touches the horizon, and ends when it disappears completely. If the Sun touches the horizon and then lifts back up, the sunset is disqualified.

For a sunset to count, the Sun has to set behind the idealized horizon, not just behind a nearby hill. This is not a sunset, even though it seems like one:

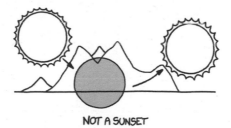

NOT A SUNSET

The reason it can't count as a sunset is that if you could use arbitrary obstacles, you could cause a sunset at any time by hiding behind a rock.

We also have to consider refraction. The Earth's atmosphere bends light, so when the Sun is at the horizon it appears about one Sun-width higher than it would otherwise. The standard practice seems to be to include the average effect of this in all calculations, which I've done here.

At the equator in March and September, sunset is a hair over two minutes long. Closer to the poles, in places like London, it can take between 200 and 300 seconds. It's shortest in spring and fall (when the Sun is over the equator) and longest in the summer and winter.

If you stand still at the South Pole in early March, the Sun stays in the sky all day, making a full circle just above the horizon. Sometime around March 21, it touches the horizon for the only sunset of the year. This sunset takes 38–40 hours, which means it makes more than a full circuit around the horizon while setting.

But Michael's question was very clever. He asked about the longest sunset you can experience on a paved road. There's a road to the research station at the South Pole, but it's not paved—it's made of packed snow. There are no paved roads anywhere near either pole.

The closest road to either pole that really qualifies as paved is probably the main road in Longyearbyen, on the island of Svalbard, Norway. (The end of the airport runway in Longyearbyen gets you slightly closer to the pole, although driving a car there might get you in trouble.)

Longyearbyen is actually closer to the North Pole than McMurdo Station in Antarctica is to the South Pole. There are a handful of military, research, and fishing stations farther north, but none of them have much in the way of roads; just airstrips, which are usually gravel and snow.

If you putter around downtown Longyearbyen,[1] the longest sunset you could experience would be a few minutes short of an hour. It doesn't actually matter if you drive or not; the town is too small for your movement to make a difference.

But if you head over to the mainland, where the roads are longer, you can do even better.

If you start driving from the tropics and stay on paved roads, the farthest north you can get is the tip of European Route 69 in Norway. There are a number of roads crisscrossing northern Scandinavia, so that seems like a good place to start. But which road should we use?

Intuitively, it seems like we want to be as far north as possible. The closer we are to the pole, the easier it is to keep up with the Sun.

---

1 Get a picture with the "polar bear crossing" sign.

Unfortunately, it turns out keeping up with the Sun isn't a good strategy. Even in those high Norwegian latitudes, the Sun is just too fast. At the tip of European Route 69—the farthest you can get from the equator while driving on paved roads—you'd still have to drive at about half the speed of sound to keep up with the Sun. (And E69 runs north-south, not east-west, so you'd drive into the Barents Sea anyway.)

Luckily, there's a better approach.

If you're in northern Norway on a day when the Sun just barely sets and then rises again, the terminator (day-night line) moves across the land in this pattern:

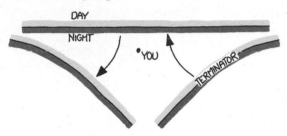

(Not to be confused with the Terminator, which moves across the land in this pattern:)

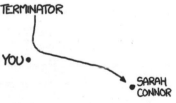

I can't decide which terminator I'd rather have to run from.

To get a long sunset, the strategy is simple: Wait for the date when the terminator will just barely reach your position. Sit in your car until the terminator reaches you, drive north to stay a little ahead of it for as long as you can (depending on the local road layout), then U-turn and drive back south fast enough that you can get past it to the safety of darkness.[2]

Surprisingly, this strategy works about equally well anywhere inside the Arc-

---

2 These instructions also work for the other kind of Terminator.

tic Circle; so you can get this lengthy sunset on many roads across Finland and Norway. I ran a search for long-sunset driving paths using PyEphem and some GPS traces of Norwegian highways. I found that over a wide range of routes and driving speeds, the longest sunset was consistently about 95 minutes—an improvement of about 40 minutes over the Svalbard sit-in-one-place strategy.

But if you are stuck in Svalbard and want to make the sunset—or sunrise—last a little longer, you can always try spinning counterclockwise.[3] It's true that it will add only an immeasurably small fraction of a nanosecond to the Earth's clock. But depending on who you're with . . .

. . . it might be worth it.

3 xkcd, "Angular Momentum," *http://xkcd.com/162/*.

Q. If you call a random phone number and say "God bless you," what are the chances that the person who answers just sneezed?

—**Mimi**

A. IT'S HARD TO FIND good numbers on this, but it's probably about 1 in 40,000.

Before you pick up the phone, you should also keep in mind that there's roughly a 1 in 1000,000,000 chance that the person you're calling just murdered someone.[1] You may want to be more careful with your blessings.

However, given that sneezes are far more common than murders,[2] you're still much more likely to get someone who sneezed than to catch a killer, so this strategy is not recommended.

Mental note: I'm going to start saying this when people sneeze.

Compared with the murder rate, the sneezing rate doesn't get much scholarly research. The most widely cited figure for average sneeze frequency comes from a doctor interviewed by ABC News, who pegged it at 200 sneezes per person per year.

One of the few scholarly sources of data on sneezing is a study that monitored the sneezing of people undergoing an induced allergic reaction. To estimate the average sneezing rate, we can ignore all the real medical data they were trying to gather and just look at their control group. This group was given no allergens at all; they just sat alone in a room for a total of 176 20-minute sessions.[3]

The subjects in the control group sneezed four times during those 58 or so hours,[4] which—assuming they sneeze only while awake—translates to about 400 sneezes per person per year.

Google Scholar turns up 5980 articles from 2012 that mention "sneezing." If half of these articles are from the US, and each one has an average of four

---

[1] Based on a rate of 4 per 100,000, which is the average in the US but on the high end for industrialized countries.

[2] Citation: You are alive.

[3] For context, that's 490 repetitions of the song "Hey Jude."

[4] Over 58 hours of research, four sneezes were the most interesting data points. I might've taken the 490 "Hey Jude"s.

authors, then if you dial the number, there's about a 1 in 10,000,000 chance that you'll get someone who—just that day—published an article on sneezing.

On the other hand, about 60 people are killed by lightning in the US every year. That means there's only a 1 in 10,000,000,000,000 chance that you'll call someone in the 30 seconds after they've been struck and killed.

Lastly, let's suppose that on the day this book was published, five people who read it decide to actually try this experiment. If they call numbers all day, there's about a 1 in 30,000 chance that at some point during the day, one of them will get a busy signal because the person they've called is also calling a random stranger to say "God bless you."

And there's about a 1 in 10,000,000,000,000 chance that two of them will simultaneously call each other.

At this point, probability will give up, and they'll both be struck by lightning.

WEIRD (AND WORRYING) QUESTIONS
FROM THE WHAT IF? INBOX, #10

Q. What is the probability that if I am stabbed by a knife in my torso that it won't hit anything vital and I'll live?

—Thomas

... ASKING FOR A FRIEND.

FORMER FRIEND, I MEAN.

Q. If I were on a motorbike and do a jump off a quarter pipe ramp, how fast would I need to be moving to safely deploy and land using the parachute?

—Anonymous

Q. What if every day, every human had a 1 percent chance of being turned into a turkey, and every turkey had a 1 percent chance of being turned into a human?

—Kenneth

Q. How long would it take for people to notice their weight gain if the mean radius of the world expanded by 1cm every second? (Assuming the average composition of rock were maintained.)

—**Dennis O'Donnell**

A. THE EARTH IS NOT, currently, expanding.

People have long suggested that it might be. Before the continental drift hypothesis was confirmed in the 1960s,[1] people had noticed that the continents fit together. Various ideas were put forward to explain this, including the idea that the ocean basins were rifts that opened in the surface of a previously smooth Earth as it expanded. This theory was never very widespread,[2] although it still periodically makes the rounds on YouTube.

----

[1] The smoking gun that confirmed the theory of plate tectonics was the discovery of sea-floor spreading. The way sea-floor spreading and magnetic pole reversal neatly confirmed each other is one of my favorite examples of scientific discovery at work.
[2] It turns out it's kind of dumb.

To avoid the problem of rifts in the ground, let's imagine all the matter in the Earth, from the crust to the core, starts expanding uniformly. To avoid another drain-the-oceans scenario, we'll assume the ocean expands, too.[3] All human structures will stay.

t = 1 second

As the Earth started expanding, you'd feel a slight jolt, and might even lose your balance for a moment. This would be very brief. Since you're moving steadily upward at 1 cm/s, you woudn't feel any kind of ongoing acceleration. For the rest of the day, you wouldn't notice much of anything.

t = 1 day

After the first day, the Earth would have expanded by 864 meters.

Gravity would take a long time to increase noticeably. If you weighed 70 kilograms when the expansion started, you'd weigh 70.01 at the end of the first day.

What about our roads and bridges? Eventually, they would have to break up, right?

Not as quickly as you might think. Here's a puzzle I once heard:

---

3 As it turns out, the ocean is expanding, since it's getting warmer. This is (currently) the main way global warming *is* raising the sea level.

Imagine you tied a rope tightly around the Earth, so it was hugging the surface all the way around.

Now imagine you wanted to raise the rope 1 meter off the ground.

How much extra length will you need to add to the rope?

Though it may seem like you'd need miles of rope, the answer is 6.28 meters. Circumference is proportional to radius, so if you increase radius by 1 unit, you increase circumference by 2π units.

Stretching a 40,000-kilometer line an extra 6.28 meters is pretty negligible. Even after a day, the extra 5.4 kilometers would be handled easily by virtually all structures. Concrete expands and contracts by more than that every day.

After the initial jolt, one of the first effects you'd notice would be that your GPS would stop working. The satellites would stay in roughly the same orbits, but the delicate timing that the GPS system is based on would be completely ruined within hours. GPS timing is incredibly precise; of all the problems in engineering, it's one of the only ones in which engineers have been forced to include both special *and* general relativity in their calculations.

Most other clocks would keep working fine. However, if you have a very precise pendulum clock, you might notice something odd—by the end of the day, it would be three seconds ahead of where it should be.

t = 1 month

After a month, the Earth would have expanded by 26 kilometers—an increase of 0.4 percent—and its mass would have increased by 1.2 percent. Surface gravity would have gone up by only 0.4 percent, rather than 1.2 percent, since surface gravity is proportional to radius.[4]

You might notice the difference in weight on a scale, but it's not a big deal. Gravity varies by this much between different cities already. This is a good thing to keep in mind if you buy a digital scale. If your scale has a precision of more than two decimal places, you need to calibrate it with a test weight—the force of gravity at the scale factory isn't necessarily the same as the force of gravity at your house.

While you might not notice the increased gravity just yet, you'd notice the expansion. After a month, you'd see a lot of cracks opening up in long concrete structures and the failure of elevated roads and old bridges. Most buildings would probably be OK, although those anchored firmly into bedrock might start to behave unpredictably.[5]

At this point, astronauts on the ISS would start getting worried. Not only would the ground (and atmosphere) be rising toward them, but the increased gravity would also cause their orbit to slowly shrink. They'd need to evacuate quickly; they'd have at most a few months before the station reentered the atmosphere and deorbited.

t = 1 year

After a year, gravity would be 5 percent stronger. You'd probably notice the weight gain, and you'd definitely notice the failure of roads, bridges, power lines, satellites, and undersea cables. Your pendulum clock would now be ahead by five days.

What about the atmosphere?

If the atmosphere isn't growing like the land and water are, air pressure would start dropping. This is due to a combination of factors. As gravity increases, then air gets heavier. But since that air is spread out over a larger area, the overall effect would be *decreasing* air pressure.

On the other hand, if the atmosphere is also expanding, surface air pressure would rise. After years had passed, the top of Mount Everest would no longer be in the "death zone." On the other hand, since you'd be heavier—and the mountain would be taller—climbing would be more work.

---

4 Mass is proportional to radius cubed, and gravity is proportional to mass times inverse square of radius, so radius³ / radius² = radius.
5 Just what you want in a skyscraper.

t = 5 years

After five years, gravity would be 25 percent stronger. If you weighed 70 kg when the expansion started, you'd weigh 88 kg now.

Most of our infrastructure would have collapsed. The cause of the collapse would be the expanding ground below them, not the increased gravity. Surprisingly, most skyscrapers would hold up fine under much higher gravity.[6] For most of them, the limiting factor isn't weight, but wind.

t = 10 years

After 10 years, gravity would be 50 percent stronger. In the scenario where the atmosphere isn't expanding, the air would become thin enough to be difficult to breathe even at sea level. In the other scenario, we'd be OK for a little while longer.

t = 40 years

After 40 years, Earth's surface gravity would have tripled.[7] At this point, even the strongest humans would be able to walk only with great difficulty. Breathing would be difficult. Trees would collapse. Crops wouldn't stand up under their own weight. Virtually every mountainside would see massive landslides as material sought out a shallower angle of repose.

Geologic activity would also accelerate. Most of the Earth's heat is provided by radioactive decay of minerals in the crust and mantle,[8] and more Earth means more heat. Since the volume expands faster than the surface area, the overall heat flowing out per square meter will increase.

It's not actually enough to substantially warm the planet—Earth's surface temperature is dominated by the atmosphere and the Sun—but it would lead to more volcanoes, more earthquakes, and faster tectonic movement. This would be similar to the situation on Earth billions of years ago, when we had more radioactive material and a hotter mantle.

---

6   Although I wouldn't trust the elevators.
7   Over decades, the force of gravity would grow slightly faster than you'd expect, since the material in the Earth would compress under its own weight. The pressure inside planets is roughly proportional to the square of their surface area, so the Earth's core would be squeezed tightly.
    http://cseligman.com/text/planets/internalpressure.htm.
8   Although some radioactive elements, like uranium, are heavy, they get squeezed out of the lower layers because their atoms don't mesh well with the rock lattices at those depths. For more, see this chapter: *http://igppweb.ucsd.edu/~guy/sio103/chap3.pdf* and this article: *http://world-nuclear.org/info/ Nuclear-Fuel-Cycle/Uranium-Resources/The-Cosmic-Origins-of-Uranium/#.UlxuGmRDJf4.*

WEIGHTLESS ARROW

Q. Assuming a zero-gravity environment with an atmosphere identical to Earth's, how long would it take the friction of air to stop an arrow fired from a bow? Would it eventually come to a standstill and hover in midair?

—**Mark Estano**

A. **IT'S HAPPENED TO ALL** of us. You're in the belly of a vast space station and you're trying to shoot someone with a bow and arrow.

Compared to a normal physics problem, this scenario is backward. Usually, you consider gravity and neglect air resistance, not the other way around.[1]

As you'd expect, air resistance would slow down an arrow, and eventually it would stop . . . after flying very, very far. Fortunately, for most of that flight, it wouldn't be much of a danger to anyone.

Let's go over what would happen in more detail.

Say you fire the arrow at 85 meters per second. That's about twice the speed of a major-league fastball, and a little below the 100 m/s speed of arrows from high-end compound bows.

The arrow would slow down quickly. Air resistance is proportional to speed squared, which means that when it's going fast, the arrow would experience a lot of drag.

After ten seconds of flight, the arrow would have traveled 400 meters, and its speed would have dropped from 85 m/s to 25 m/s; 25 m/s is about how fast a normal person could *throw* an arrow.

UH, LEGOLAS?

At that speed, the arrow would be a lot less dangerous.

We know from hunters that small differences in arrow speed make big differences in the size of the animal it can kill. A 25-gram arrow moving at 100 m/s could be used to hunt elk and black bears. At 70 m/s, it might be too slow to kill a deer. Or, in our case, a space deer.

Once the arrow leaves that range, it's no longer particularly dangerous . . . but it's not even close to stopping.

After five minutes, the arrow would have flown about a mile, and it would have slowed to roughly walking speed. At that speed, it would experience very little drag; it would just cruise along, slowing down very gradually.

---

[1] Also, you don't usually shoot astronauts with a bow and arrow—at least not for an undergraduate degree.

At this point, it would have gone much farther than any Earth arrow can go. High-end bows can shoot an arrow a distance of a couple hundred meters over flat ground, but the world record for a hand bow-and-arrow shot is just over a kilometer.

This record was set in 1987 by archer Don Brown. Brown set his record by firing slender metal rods from a terrifying contraption that only vaguely resembled a traditional bow.

As the minutes stretch into hours and the arrow slows down more and more, the airflow changes.

Air has very little viscosity. That is, it's not gooey. That means things flying through the air experience drag because of the momentum of the air they're shoving out of the way—not from cohesion between the air molecules. It's more like pushing your hand through a bathtub full of water than a bathtub full of honey.

After a few hours, the arrow would be moving so slowly that it would be barely visible. At this point, assuming the air is relatively still, the air would start acting like honey instead of water. And the arrow would, very gradually, come to a stop.

The exact range would depend heavily on the precise design of the arrow. Small differences in an arrow's shape can dramatically change the nature of the airflow over it at low speeds. But at minimum, it would probably fly several kilometers, and could conceivably go as far as 5 or 10.

Here's the problem: Currently, the only sustained zero-g environment with an Earth-like atmosphere is the International Space Station. And the largest ISS module, Kibo, is only 10 meters long.

This means that if you actually performed this experiment, the arrow would fly no more than 10 meters. Then, it would either come to a stop . . . or *really* ruin someone's day.

SUNLESS EARTH

Q. What would happen to the Earth if the Sun suddenly switched off?

—Many, many readers

A. THIS IS PROBABLY THE single most popular submission to *What If*.

Part of why I haven't answered it is that it's been answered already. A Google search for "what if the Sun went out" turns up a lot of excellent articles thoroughly analyzing the situation.

However, the rate of submission of this question continues to rise, so I've decided to do my best to answer it.

If the Sun went out . . .

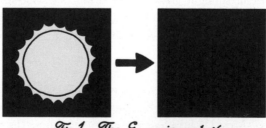

Fig. 1. The Sun going out :(

We won't worry about exactly how it happens. We'll just assume we figured out a way to fast-forward the Sun through its evolution so that it becomes a cold, inert sphere. What would the consequences be for us here on Earth?

Let's look at a few . . .

Reduced risk of solar flares: In 1859, a massive solar flare and geomagnetic storm hit the Earth. Magnetic storms induce electric currents in wires. Unfortunately for us, by 1859 we had wrapped the Earth in telegraph wires. The storm caused powerful currents in those wires, knocking out communications and in some cases causing telegraph equipment to catch fire.

Since 1859, we've wrapped the Earth in a lot more wires. If the 1859 storm hit us today, the Department of Homeland Security estimates the economic damage to the US alone would be several trillion dollars—more than every hurricane that has ever hit the US *combined*. If the Sun went out, this threat would be eliminated.

Improved satellite service: When a communications satellite passes in front of the Sun, the Sun can drown out the satellite's radio signal, causing an interruption in service. Deactivating the Sun would solve this problem.

Better astronomy: Without the Sun, ground-based observatories would be able to operate around the clock. The cooler air would create less atmospheric noise, which would reduce the load on adaptive optics systems and allow for sharper images.

Stable dust: Without sunlight, there would be no Poynting–Robertson drag, which means we would finally be able to place dust into a stable orbit around the Sun without the orbits decaying. I'm not sure whether anyone wants to do that, but you never know.

Reduced infrastructure costs: The Department of Transportation estimates that it would cost $20 billion per year over the next 20 years to repair and maintain all US bridges. Most US bridges are over water; without the Sun, we could save money by simply driving on a strip of asphalt laid across the ice.

Cheaper trade: Time zones make trade more expensive; it's harder to do business with someone if their office hours don't overlap with yours. If the Sun went out, it would eliminate the need for time zones, allowing us to switch to UTC and give a boost to the global economy.

Safer children: According to the North Dakota Department of Health, babies younger than six months should be kept out of direct sunlight. Without sunlight, our children would be safer.

Safer combat pilots: Many people sneeze when exposed to bright sunlight. The reasons for this reflex are unknown, and it may pose a danger to fighter pilots during flight. If the Sun went dark, it would mitigate this danger to our pilots.

Safer parsnip: Wild parsnip is a surprisingly nasty plant. Its leaves contain chemicals called furocoumarins, which can be absorbed by human skin without causing symptoms . . . at first. However, when the skin is then exposed to sunlight (even days or weeks later), the furocoumarins cause a nasty chemical burn. This is called phytophotodermatitis. A darkened Sun would liberate us from the parsnip threat.

HIKING TIP:
WHAT TO DO IF YOU COME ACROSS WILD PARSNIP:

In conclusion, if the Sun went out, we would see a variety of benefits across many areas of our lives.

Are there any downsides to this scenario?

We would all freeze and die.

UPDATING A PRINTED WIKIPEDIA

Q. If you had a printed version of the whole of (say, the English) Wikipedia, how many printers would you need in order to keep up with the changes made to the live version?

—Marein Könings

A. THIS MANY.

If a date took you home and you saw a row of working printers set up in his or her living room, what would you think?

That's surprisingly few printers! But before you try to create a live-updating paper Wikipedia, let's look at what those printers would be *doing* . . . and how much they'd cost.

Printing Wikipedia

People have considered printing out Wikipedia before. One student, Rob Matthews, printed every Wikipedia featured article, creating a book several feet thick.

Of course, that's just a small slice of the best of Wikipedia; the entire encyclopedia would be a lot bigger. Wikipedia user **Tompw** has set up a tool that calculates the current size of the whole English Wikipedia in printed volumes. It would fill a lot of bookshelves.

Keeping up with the edits would be hard.

Keeping up

The English Wikipedia currently receives about 125,000 to 150,000 edits each day, or 90–100 per minute.

We could try to define a way to measure the "word count" of the average edit, but that's hard bordering on impossible. Fortunately, we don't need to—we can just estimate that each change is going to require us to reprint a page somewhere. Many edits will actually change multiple pages—but many other edits are reverts, which would let us put back pages we've already printed.[1] One page per edit seems like a reasonable middle ground.

For a mix of photos, tables, and text typical of Wikipedia, a good inkjet printer might put out 15 pages per minute. That means you'd need only about six printers running at any given time to keep pace with the edits.

The paper would stack up quickly. Using Rob Matthews' book as a starting point, I did my own back-of-the-envelope estimate for the size of the current English Wikipedia. Based on the average length of featured articles vs. all articles, I came up with an estimate of 300 cubic meters for a printout of the whole thing in plain text form.

By comparison, if you were trying to keep up with the edits, you'd print out 300 cubic meters every *month*.

$500,000 per month

Six printers isn't that many, but they'd be running all the time. And that gets expensive.

The electricity to run them would be cheap—a few dollars a day.

---

[1] The filing system that would be required for this would be mind-bending. I'm fighting the urge to start trying to design it.

The paper would be about 1 cent per sheet, which means you'll be spending about a thousand dollars a day on paper. You'd need to hire people to manage the printers 24/7, but that would actually cost less than the paper.

Even the printers themselves wouldn't be too expensive, despite the terrifyingly fast replacement cycle.

But the *ink* cartridges would be a nightmare.

Ink

A study by QualityLogic found that for a typical inkjet printer, the real-life cost of ink ran from 5 cents per page for black-and-white to around 30 cents per page for photos. That means you'd be spending four to five figures per *day* on ink cartridges.

You definitely want to invest in a laser printer. Otherwise, in just a month or two, this project could end up costing you half a million dollars:

But that's not even the worst part.

On January 18, 2012, Wikipedia blacked out all its pages to protest proposed Internet-freedom-limiting laws. If, someday, Wikipedia decides to go dark again, and you want to join the protest . . .

. . . . you'll have to get a crate of markers and color every page solid black yourself.

I would definitely stick to digital.

Q. When, if ever, will Facebook contain more profiles of dead people than of living ones?

—Emily Dunham

"Put on your headphones!" "Can't. Ears fell off."

A. EITHER THE 2060s or the 2130s.

There are not a lot of dead people on Facebook.[1] The main reasons for this is that Facebook—and its users—are young. The average Facebook user has gotten older over the last few years, but the site is still used at a much higher rate by the young than by the old.

The past

Based on the site's growth rate, and the age breakdown of its users over time,[2] there are probably 10 to 20 million people who created Facebook profiles who have since died.

---

1 At the time I wrote this, anyway, which was before the bloody robot revolution.
2 You can get user counts for each age group from Facebook's create-an-ad tool, although you may want to try to account for the fact that Facebook's age limits cause some people to lie about their ages.

These people are, at the moment, spread out pretty evenly across the age spectrum. Young people have a much lower death rate than people in their 60s or 70s, but they make up a substantial share of the dead on Facebook simply because there have been so many of them using it.

An elderly Cory Doctorow cosplaying by wearing what the future thinks he wore in the past.

The future

About 290,000 US Facebook users probably died in 2013. The worldwide total for 2013 is likely several million.[3] In just seven years, this death rate will double, and in seven more years it will double again.

Even if Facebook closes registration tomorrow, the number of deaths per year will continue to grow for many decades, as the generation who was in college between 2000 and 2020 grows old.

The deciding factor in when the dead will outnumber the living is whether Facebook adds new living users—ideally, young ones—fast enough to outrun this tide of death for a while.

Facebook 2100

This brings us to the question of Facebook's future.

We don't have enough experience with social networks to say with any kind of certainty how long Facebook will last. Most websites have flared up and then gradually declined in popularity, so it's reasonable to assume Facebook will follow that pattern.[4]

In that scenario, where Facebook starts losing market share later this decade and never recovers, Facebook's crossover date—the date when the dead outnumber the living—will come sometime around 2065.

---

3 Note: In some of these projections, I used US age/usage data extrapolated to the Facebook userbase as a whole, because it's easier to find US census and actuarial numbers than to assemble the country-by-country for the whole Facebook-using world. The US isn't a perfect model of the world, but the basic dynamics—young people's Facebook adoption determines the site's success or failure while population growth continues for a while and then levels off—will probably hold approximately true. If we assume a rapid Facebook saturation in the developing world, which currently has a faster-growing and younger population, it shifts many of the landmarks by a handful of years, but doesn't change the overall picture as much as you might expect.

4 I'm assuming, in these cases, that no data is ever deleted. So far, that's been a reasonable assumption; if you've made a Facebook profile, that data probably still exists, and most people who stop using a service don't bother to delete their profiles. If that behavior changes, or if Facebook performs a mass purging of its archives, the balance could change rapidly and unpredictably.

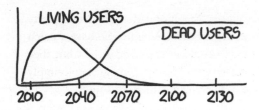

But maybe it won't. Maybe it will take on a role like the TCP protocol, where it becomes a piece of infrastructure on which other things are built, and has the inertia of consensus.

If Facebook is with us for generations, then the crossover date could be as late as the mid-2100s.

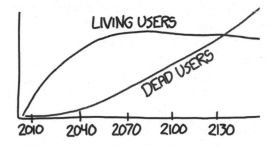

That seems unlikely. Nothing lasts forever, and rapid change has been the norm for anything built on computer technology. The ground is littered with the bones of websites and technologies that seemed like permanent institutions ten years ago.

It's possible the reality could be somewhere in between.[5] We'll just have to wait and find out.

The fate of our accounts

Facebook can afford to keep all our pages and data indefinitely. Living users will always generate more data than dead ones,[6] and the accounts for active users are the ones that will need to be easily accessible. Even if accounts for dead (or inactive) people make up a majority of their users, it will probably never add up to a large part of its overall infrastructure budget.

---

5 Of course, if there's a sudden rapid increase in the death rate of Facebook users—possibly one that includes humans in general—the crossover could happen tomorrow.

6 I hope.

More important will be our decisions. What do we *want* for those pages? Unless we demand that Facebook deletes them, they will presumably, by default, keep copies of everything forever. Even if they don't, other data-vacuuming organizations will.

Right now, next of kin can convert a dead person's Facebook profile into a memorial page. But there are a lot of questions surrounding passwords and access to private data that we haven't yet developed social norms for. Should accounts remain accessible? What should be made private? Should next of kin have the right to access email? Should memorial pages have comments? How do we handle trolling and vandalism? Should people be allowed to interact with dead user accounts? What lists of friends should they show up on?

These are issues that we're currently in the process of sorting out by trial and error. Death has always been a big, difficult, and emotionally charged subject, and every society finds different ways to handle it.

The basic pieces that make up a human life don't change. We've always eaten, learned, grown, fallen in love, fought, and died. In every place, culture, and technological landscape, we develop a different set of behaviors around these same activites.

Like every group that came before us, we're learning how to play those same games on our particular playing field. We're developing, through sometimes messy trial and error, a new set of social norms for dating, arguing, learning, and growing on the Internet. Sooner or later, we'll figure out how to mourn.

Q. When (if ever) did the Sun finally set on the British Empire?

—Kurt Amundson

A. IT HASN'T. YET. BUT only because of a few dozen people living in an area smaller than Disney World.

The world's largest empire

The British Empire spanned the globe. This led to the saying that the Sun never set on it, since it was always daytime somewhere in the Empire.

It's hard to figure out exactly when this long daylight began. The whole process of claiming a colony (on land already occupied by other people) is awfully arbitrary in the first place. Essentially, the British built their empire by sailing around and sticking flags on random beaches. This makes it hard to decide when a particular spot in a country was "officially" added to the Empire.

"What about that shadowy place over there?" "That's France. We'll get it one of these days."

The exact day when the Sun stopped setting on the Empire was probably sometime in the late 1700s or early 1800s, when the first Australian territories were added.

The Empire largely disintegrated in the early 20th century, but—surprisingly—the Sun hasn't technically started setting on it again.

Fourteen territories

Britain has 14 overseas territories, the direct remnants of the British Empire.

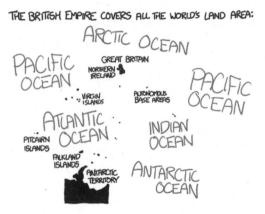

Many newly independent British colonies joined the Commonwealth of Nations. Some of them, like Canada and Australia, have Queen Elizabeth as their official monarch. However, they are independent states that happen to have the same queen; they are not part of any empire.[1]

The Sun never sets on all 14 British territories at once (or even 13, if you don't count the British Antarctic Territory). However, if the UK loses one tiny territory, it will experience its first Empire-wide sunset in over two centuries.

Every night, around midnight GMT, the Sun sets on the Cayman Islands, and doesn't rise over the British Indian Ocean Territory until after 1:00 A.M. For that hour, the little Pitcairn Islands in the South Pacific are the only British territory in the Sun.

The Pitcairn Islands have a population of a few dozen people, the descendants of the mutineers from the HMS *Bounty*. The islands became notorious

---

1   That they know of.

in 2004 when a third of the adult male population, including the mayor, were convicted of child sexual abuse.

As awful as the islands may be, they remain part of the British Empire, and unless they're kicked out, the two-century-long British daylight will continue.

Will it last *forever?*

Well, maybe.

In April of 2432, the island will experience its first total solar eclipse since the mutineers arrived.

Luckily for the Empire, the eclipse happens at a time when the Sun is over the Cayman Islands in the Caribbean. Those areas won't see a total eclipse; the Sun will even still be shining in London.

In fact, no total eclipse for the next thousand years will pass over the Pitcairn Islands at the right time of day to end the streak. If the UK keeps its current territories and borders, it can stretch out the daylight for a long, long time.

But not forever. Eventually—many millennia in the future—an eclipse will come for the island, and the Sun will finally set on the British Empire.

Q. I was absentmindedly stirring a cup of hot tea, when I got to thinking, "Aren't I actually adding kinetic energy into this cup?" I know that stirring does help to cool down the tea, but what if I were to stir it faster? Would I be able to boil a cup of water by stirring?

—**Will Evans**

A. NO.

The basic idea makes sense. Temperature is just kinetic energy. When you stir tea, you're adding kinetic energy to it, and that energy goes somewhere. Since the tea doesn't do anything dramatic like rise into the air or emit light, the energy must be turning to heat.

Am I making tea wrong?

The reason you don't notice the heat is that you're not adding very much of it. It takes a huge amount of energy to heat water; by volume, it has a greater heat capacity than any other common substance.[1]

If you want to heat water from room temperature to nearly boiling in two minutes, you'll need a lot of power:[2]

$$1 \text{ cup} \times \text{Water heat capacity} \times \frac{100°\text{C-}20°\text{C}}{2 \text{ minutes}} = 700 \text{ watts}$$

Our formula tells us that if we want to make a cup of hot water in two minutes, we'll need a 700-watt power source. A typical microwave uses 700 to 1100 watts, and it takes about two minutes to heat a mug of water to make tea. It's nice when things work out![3]

Microwaving a cup of water for two minutes at 700 watts delivers an awful lot of energy to the water. When water falls from the top of Niagara Falls, it gains kinetic energy, which is converted to heat at the bottom. But even after falling that great distance, the water heats up by only a fraction of a degree.[4] To boil a cup of water, you'd have to drop it from higher than the top of the atmosphere.

---

1 Hydrogen and helium have a higher heat capacity by mass, but they're diffuse gasses. The only other common substance with a higher heat capacity by mass is ammonia. All three of these lose to water when measured by volume.

2 Note: Pushing almost-boiling water to boiling takes a large burst of extra energy on top of what's required to heat it to the boiling point—this is called the enthalpy of vaporization.

3 If they didn't, we'd just blame "inefficiency" or "vortices."

4 $\text{Height of Niagra Falls} \times \dfrac{\text{Acceleration of gravity}}{\text{Specific heat of water}} = 0.12°\text{C}$

(The British Felix Baumgartner)

How does stirring compare to microwaving?

Based on figures from industrial mixer engineering reports, I estimate that vigorously stirring a cup of tea adds heat at a rate of about a ten-millionth of a watt. That's completely negligible.

The physical effect of stirring is actually a little complicated.[5] Most of the heat is carried away from teacups by the air convecting over them, and so they cool from the top down. Stirring brings fresh hot water from the depths, so it can help this process. But there are other things going on—stirring disturbs the air, and it heats the walls of the mug. It's hard to be sure what's really going on without data.

Fortunately, we have the Internet. Stack Exchange user **drhodes** measured the rate of teacup cooling from stirring vs. not stirring vs. repeatedly dipping a spoon into the cup vs. lifting it. Helpfully, **drhodes** posted both high-resolution graphs *and* the raw data itself, which is more than you can say for a lot of journal articles.

The conclusion: It doesn't really matter whether you stir, dip, or do nothing; the tea cools at about the same rate (although dipping the spoon in and out of the tea cooled it slightly faster).

---

5   In some situations, mixing liquids can actually help keep them warm. Hot water rises, and when a body of water is large and still enough (like the ocean), a warm layer forms on the surface. This warm layer radiates heat much more quickly than a cold layer would. If you disrupt this hot layer by mixing the water, the rate of heat loss decreases.
    This is why hurricanes tend to lose strength if they stop moving forward—their waves churn up cold water from the depths, cutting them off from the thin layer of hot surface water that was their main source of energy.

Which brings us back to the original question: Could you boil tea if you just stirred it hard enough?

No.

The first problem is power. The amount of power in question, 700 watts, is about a horsepower, so if you want to boil tea in two minutes, you'll need at least one horse to stir it hard enough.

You can reduce the power requirement by heating the tea over a longer period of time, but if you reduce it too far the tea will be cooling as fast as you're heating it.

Even if you could churn the spoon hard enough—tens of thousands of stirs per second—fluid dynamics would get in the way. At those high speeds, the tea would cavitate; a vacuum would form along the path of the spoon and stirring would become ineffective.[6]

And if you stir hard enough that your tea cavitates, its surface area will increase very rapidly, and it will cool to room temperature in seconds.

No matter how hard you stir your tea, it's not going to get any warmer.

---

6 Some blenders, which are enclosed, actually do manage to warm their contents this way. But what kind of person makes tea in a *blender*?

ALL THE LIGHTNING

Q. If all the lightning strikes happening in the world on any given day all happened in the same place at once, what would happen to that place?

—Trevor Jones

A. THEY SAY LIGHTNING NEVER strikes in the same place twice.

"They" are wrong. From an evolutionary perspective, it's a little surprising that this saying has survived; you'd think that people who believed it would have been gradually filtered out of the living population.

This is how evolution works, right?

People often wonder whether we could harvest electrical power from lightning. On the face of it, it makes sense; after all, lightning is electricity,[1] and there is indeed a substantial amount of power in a lightning strike. The problem is, it's hard to get lightning to strike where you want it.[2]

A typical lightning strike delivers enough energy to power a residential house for about two days. That means that even the Empire State Building, which is struck by lightning about 100 times a year, wouldn't be able to keep a house running on lightning power alone.

Even in regions of the world with a lot of lightning, such as Florida and the eastern Congo, the power delivered to the ground by sunlight outweighs the power delivered by lightning by a factor of a million. Generating power from lightning is like building a wind farm whose blades are turned by a tornado: ~~awesome~~ impractical.[3]

Trevor's lightning

In Trevor's scenario, all the lightning in the world hits in one place. This would make power generation a lot more attractive!

By "happened in the same place," let's assume the lightning bolts all come down in parallel, right up against each other. The main channel of a lightning bolt—the part that's carrying current—is about a centimeter in diameter. Our bundle contains about a million separate bolts, which means it will be about 6 meters in diameter.

Every science writer always compares everything to the atomic bomb dropped on Hiroshima,[4] so we may as well get that out of the way: The lightning bolt would deliver about two atomic bombs' worth of energy to the air and ground.

1 Citation: The presentation I gave to my third-grade class at Assawompset Elementary School while wearing a Ben Franklin costume.
2 And I hear it never strikes in the same place twice.
3 In case you're curious, yes, I did run some numbers on using passing tornadoes to run wind turbines, and it's even less practical than gathering lightning. The average location in the heart of Tornado Alley has a tornado pass over it only every 4000 years. Even if you managed to absorb all the accumulated energy of the tornado, it would still result in less than a watt of average power output in the long run. Believe it or not, something like this idea has actually been attempted. A company called AVEtec has proposed building a "vortex engine" that would produce artificial tornadoes and use them to generate power.
4 Niagara Falls has a power output equal to a Hiroshima-sized bomb going off **every eight hours**! The atomic bomb dropped on Nagasaki had an explosive power equal to **1.3 Hiroshima bombs**! For context, the gentle breeze blowing across a prairie *also* carries roughly the kinetic energy of a Hiroshima bomb.

From a more practical standpoint, this is enough electricity to power a game console and plasma TV for several million years. Or, to put it another way, it could support the US's electricity consumption . . . for five minutes.

The bolt itself wouldn't be much wider than the center circle of a basketball court, but it would leave a crater the size of the entire court.

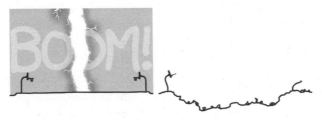

Within the bolt, the air would turn to high-energy plasma. The light and heat from the bolt would spontaneously ignite surfaces for miles around. The shockwave would flatten trees and demolish buildings. All in all, the Hiroshima comparison is not far off.

Could we protect ourselves?

Lightning rods

The mechanism by which lightning rods work is disputed. Some people claim they actually ward off lightning strikes by "bleeding" charge from the ground to the air, lowering the cloud-to-ground voltage potential and reducing the probability of a strike. The National Fire Protection Association does not currently endorse this idea.

I'm not sure what the NFPA would say about Trevor's massive lightning bolt, but a lightning rod wouldn't protect you from it. A copper cable a meter in diameter could, in theory, conduct the brief surge of current from the bolt without melting. Unfortunately, when the bolt reached the bottom end of the rod, the *ground* wouldn't conduct it so well, and the explosion of molten rock would demolish your house all the same.[5]

WHAT IF WE TRIED
LESS POWER?

_____

5 Your house would already be catching fire anyway, thanks to the thermal radiation from plasma in the air.

Catatumbo lightning

Collecting all the world's lightning into one place is obviously impossible. What about gathering all the lightning from just one area?

No place on Earth has *constant* lightning, but there's an area in Venezuela that comes close. Near the southwestern edge of Lake Maracaibo, there's a strange phenomenon: perpetual nighttime thunderstorms. There are two spots, one over the lake and one over land to the west, where thunderstorms form almost every night. These storms can generate a flash of lightning every two seconds, making Lake Maracaibo the lightning capital of the world.

If you somehow managed to channel all the bolts from a single night of Catatumbo thunderstorms down through a single lightning rod, and used it to charge a massive capacitor, it would store up enough power to run a game console and plasma TV for roughly a century.[6]

Of course, if this happened, the old saying would need even *more* revision.

---

6 Since there's no cellular data coverage on the southwest shore of Lake Maracaibo, you'll have to buy service through a satellite provider, which generally means hundreds of milliseconds of lag.

Q. What is the farthest one human being has ever been from every other living person? Were they lonely?

—Bryan J McCarter

A. IT'S HARD TO KNOW for sure!

The most likely suspects are the six Apollo command module pilots who stayed in lunar orbit during a Moon landing: Mike Collins, Dick Gordon, Stu Roosa, Al Worden, Ken Mattingly, and Ron Evans.

Each of these astronauts stayed alone in the command module while two other astronauts landed on the Moon. At the highest point in their orbit, they were about 3585 kilometers from their fellow astronauts.

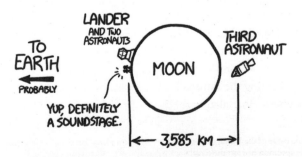

From another point of view, this was the farthest the rest of humanity has ever managed to get from those jerk astronauts.

You'd think astronauts would have a lock on this category, but it's not so cut-and-dried. There are a few other candidates who come pretty close!

Polynesians

It's hard to get 3585 kilometers from a permanently inhabited place.[1] The Polynesians, who were the first humans to spread across the Pacific, might have managed it, but this would have required a lone sailor to travel awfully far ahead of everyone else. It may have happened—perhaps by accident, when someone was carried far from their group by a storm—but we're unlikely to ever know for sure.

Once the Pacific was colonized, it got a lot harder to find regions of the Earth's surface where someone could achieve 3585-kilometer isolation. Now that the Antarctic continent has a permanent population of researchers, it's almost certainly impossible.

Antarctic explorers

During the period of Antarctic exploration, a few people have come close to beating the astronauts, and it's possible one of them actually holds the record. One person who came very close was Robert Scott.

Robert Falcon Scott was a British explorer who met a tragic end. Scott's expedition reached the South Pole in 1911, only to discover that Norwegian explorer Roald Amundsen had beaten him there by several months. The dejected Scott and his companions began their trek back to the coast, but they all died while crossing the Ross Ice Shelf.

The last surviving expedition member would have been, briefly, one of the most isolated people on Earth.[2] However, he (whoever he was) was still within 3585 kilometers of a number of humans, including some other Antarctic explorer outposts as well as the Māori on Rakiura (Stewart Island) in New Zealand.

There are plenty of other candidates. Pierre François Péron, a French sailor, says he was marooned on Île Amsterdam in the southern Indian Ocean. If so, he came close to beating the astronauts, but he wasn't quite far enough from Mauritius, southwestern Australia, or the edge of Madagascar to qualify.

---

1 Because of the curve of the Earth, you actually have to go 3619 kilometers across the surface to qualify.
2 Amundsen's expedition had left the continent by then.

We'll probably never know for sure. It's possible that some shipwrecked 18th-century sailor drifting in a lifeboat in the Southern Ocean holds the title of most isolated human. However, until some clear piece of historic evidence pops up, I think the six Apollo astronauts have a pretty good claim.

Which brings us to the second part of Bryan's question: Were they lonely?

Loneliness

After returning to Earth, Apollo 11 command module pilot Mike Collins said he did not feel at all lonely. He wrote about the experience in his book *Carrying the Fire: An Astronaut's Journeys:*

> *Far from feeling lonely or abandoned, I feel very much a part of what is taking place on the lunar surface . . . I don't mean to deny a feeling of solitude. It is there, reinforced by the fact that radio contact with the Earth abruptly cuts off at the instant I disappear behind the moon.*
>
> *I am alone now, truly alone, and absolutely isolated from any known life. I am it. If a count were taken, the score would be three billion plus two over on the other side of the moon, and one plus God knows what on this side.*

Al Worden, the Apollo 15 command module pilot, even enjoyed the experience.

> *There's a thing about being alone and there's a thing about being lonely, and they're two different things. I was alone but I was not lonely. My background was as a fighter pilot in the air force, then as a test pilot—and that was mostly in fighter airplanes—so I was very used to being by myself. I thoroughly enjoyed it. I didn't have to talk to Dave and Jim any more . . . On the backside of the Moon, I didn't even have to talk to Houston and that was the best part of the flight.*

Introverts understand; the loneliest human in history was just happy to have a few minutes of peace and quiet.

WEIRD (AND WORRYING) QUESTIONS
FROM THE WHAT IF? INBOX, #11

Q. What if everyone in Great Britain went to one of
the coasts and started paddling?
Could they move the island at all?

—Ellen Eubanks

NO.

WAIT, MAYBE WE
NEED TO DISCONNECT
THE CHUNNEL FIRST.

Q. Are fire tornadoes possible?

—Seth Wishman

YES.

FIRE TORNADOES ARE A REAL
THING THAT ACTUALLY HAPPENS.

NOTHING I SAY COULD
POSSIBLY ADD TO THIS.

Q. What if a rainstorm dropped all of its water in a single giant drop?

—Michael McNeill

A. IT'S MIDSUMMER IN KANSAS. The air is hot and heavy. Two old-timers sit on the porch in rocking chairs.

On the horizon to the southwest, ominous-looking clouds begin to appear. The towers build as they draw closer, the tops spreading out into an anvil shape.

They hear the tinkling of wind chimes as a gentle breeze picks up. The sky begins to darken.

Moisture

Air holds water. If you walled off a column of air, from the ground up to the top of the atmosphere, and then cooled the column of air down, the moisture it contained would condense out as rain. If you collected the rain in the bottom of the column, it would fill it to a depth of anywhere between zero and a few dozen centimeters. That depth is what we call the air's **total precipitable water** (TPW).

Normally, the TPW is 1 or 2 centimeters.

Satellites measure this water vapor content for every point on the globe, producing some truly beautiful maps.

We'll imagine our storm measures 100 kilometers on each side and has a high TPW content of 6 centimeters. This means the water in our rainstorm would have a volume of:

$$100\text{km} \times 100\text{km} \times 6\text{cm} = 0.6\text{km}^3$$

That water would weigh 600 million tons (which happens to be about the current weight of our species). Normally, a portion of this water would fall, scattered, as rain—at most, 6 centimeters of it.

In this storm, all that water instead condenses into one giant drop, a sphere of water over a kilometer in diameter. We'll assume it forms a couple of kilometers above the surface, since that's where most rain condenses.

The drop begins to fall.

For five or six seconds, nothing is visible. Then, the base of the cloud begins to bulge downward. For a moment, it looks a little like a funnel cloud is forming. Then the bulge widens, and at the ten-second mark, the bottom of the drop emerges from the cloud.

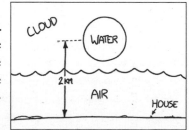

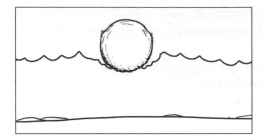

The drop is now falling at 90 meters per second (200 mph). The roaring wind whips up the surface of the water into spray. The leading edge of the droplet turns to foam as air is forced into the liquid. If it kept falling for long enough, these forces would gradually disperse the entire droplet into rain.

Before that can happen, about 20 seconds after formation, the edge of the droplet hits the ground. The water is now moving at over 200 m/s (450 mph). Right under the point of impact, the air is unable to rush out of the way fast enough, and the compression heats it so quickly that the grass would catch fire if it had time.

Fortunately for the grass, this heat lasts only a few milliseconds because it's doused by the arrival of a lot of cold water. Unfortunately for the grass, the cold water is moving at over half the speed of sound.

If you were floating in the center of this sphere during this episode, you wouldn't have felt anything unusual up until now. It'd be pretty dark in the middle, but if you had enough time (and lung capacity) to swim a few hundred meters out toward the edge, you'd be able to make out the dim glow of daylight.

As the raindrop approached the ground, the buildup of air resistance would lead to an increase in pressure that would make your ears pop. But seconds later, when the water contacted the surface, you'd be crushed to death—the shock-wave would briefly create pressures exceeding those at the bottom of the Mariana Trench.

The water plows into the ground, but the bedrock is unyielding. The pressure forces the water sideways, creating a supersonic omnidirectional jet[1] that destroys everything in its path.

The wall of water expands outward kilometer by kilometer, ripping up trees, houses, and topsoil as it goes. The house, porch, and old-timers are obliterated in an instant. Everything within a few kilometers is completely scoured away, leaving a pool of mud atop bedrock. The splash continues outward, demolishing all structures out to distances of 20 or 30 kilometers. At this distance, areas shielded by mountains or ridges are protected, and the flood begins to flow along natural valleys and waterways.

The broader region is largely protected from the effects of the storm, though areas hundreds of kilometers downstream experience flash flooding in the hours after the impact.

News trickles out into the world about the inexplicable disaster. There is widespread shock and puzzlement, and for a while, every new cloud in the sky causes mass panic. Fear reigns supreme as the world fears rain supreme, but years pass without any signs of the disaster repeating.

Atmospheric scientists try for years to piece together what happened, but no explanation is forthcoming. Eventually, they give up, and the unexplained meteorological phenomenon is simply called a "dubstep storm," because—in the words of one researcher—"It had one hell of a drop."

---

[1] Just about the coolest triplet of words I've ever seen.

Q. What if everyone who took the SAT guessed on every multiple-choice question? How many perfect scores would there be?

—Rob Balder

A. NONE.

The SAT is a standardized test given to American high school students. The scoring is such that under certain circumstances, guessing an answer can be a good strategy. But what if you *guessed* on everything?

Not all of the SAT is multiple-choice, so let's focus on the multiple-choice questions to keep things simple. We'll assume everyone gets the essay questions and fill-in-the-number sections correct.

In the 2014 version of the SAT, there were 44 multiple-choice questions in the math (quantitative) section, 67 in the critical reading (qualitative) section, and 47 in the newfangled[1] writing section. Each question has five options, so a random guess has a 20 percent chance of being right.

----

[1] I took the SAT a long time ago, okay?

The probability of getting all 158 questions right is:

$$\frac{1}{5^{44}} \times \frac{1}{5^{67}} \times \frac{1}{5^{47}} \approx \frac{1}{2.7 \times 10^{110}}$$

That's one in 27 quinquatrigintillion.

If all four million 17-year-olds took the SAT, and they all guessed randomly, it's a virtually certain that there would be no perfect scores on any of the three sections.

How certain is it? Well, if they each used a computer to take the test a million times each day, and continued this every day for five billion years—until the Sun expanded to a red giant and the Earth was charred to a cinder—the chance of any of them ever getting a perfect score on just the math section would be about 0.0001 percent.

How unlikely is that? Each year something like 500 Americans are struck by lightning (based on an average of 45 lightning deaths and a 9–10 percent fatality rate). This suggests that the odds of any one American being hit in a given year are about 1 in 700,000.[2]

This means that the odds of acing the SAT by guessing are worse than the odds of every living ex-President and every member of the main cast of *Firefly* all being independently struck by lightning . . . on the same day.

To everyone taking the SAT this year, good luck—but it won't be enough.

---

2 See: xkcd, "Conditional Risk," *http://xkcd.com/795/*.

Q. If a bullet with the density of a neutron star were fired from a handgun (ignoring the how) at the Earth's surface, would the Earth be destroyed?

—**Charlotte Ainsworth**

A. A BULLET WITH THE density of a neutron star would weigh about as much as the Empire State Building.

Whether we fired it from a gun or not, the bullet would fall straight through the ground, punching through the crust as if the rock were wet tissue paper.

We'll look at two different questions:

- What would the bullet's passage do to the Earth?
- If we kept the bullet here on the surface, what would it do to its surroundings? Could we touch it?

First, a little bit of background:

What are neutron stars?

A neutron star is what's left over after a giant star collapses under its own gravity.

Stars exist in a balance. Their massive gravity is always trying to make them collapse inward, but that squeezing sets off several different forces that push them back apart.

In the Sun, the thing holding off collapse is heat from nuclear fusion. When a star runs out of fusion fuel, it contracts (in a complicated process involving several explosions) until the collapse is stopped by the quantum laws that keep matter from overlapping with other matter.[1]

If the star is heavy enough, it overcomes that quantum pressure and collapses further (with another, more massive explosion) to become a neutron star. If the remnant is even heavier, it becomes a black hole.[2]

Neutron stars are some of the densest objects you can find (outside of the infinite density of a black hole). They're crushed by their own immense gravity into a compact quantum-mechanical soup that's in some ways similar to an atomic nucleus the size of a mountain.

Is our bullet made from a neutron star?

No. Charlotte asked for a bullet *as dense* as a neutron star, not one made from actual neutron star material. That's good, because you can't make a bullet from that stuff. If you take neutron star material outside of the crushing gravity well where it's normally found, it will re-expand into superhot normal matter with an outpouring of energy more powerful than any nuclear weapon.

That's presumably why Charlotte suggested we make our bullet out of some magical, stable material that's *as dense* as a neutron star.

What would the bullet do to the Earth?

You could imagine firing it from a gun,[3] but it might be more interesting to simply drop it. In either case, the bullet would accelerate downward, punch into the ground, and burrow toward the center of the Earth.

This wouldn't destroy the Earth, but it would be pretty strange.

As the bullet got within a few feet of the ground, the force of its gravity would yank up a huge clump of dirt, which would ripple wildly around the bullet as it

---

[1] The Pauli exclusion principle keeps electrons from getting too close to each other. This effect is one of the main reasons that your laptop doesn't fall through your lap.

[2] It's possible there's a category of objects heavier than neutron stars—but not quite heavy enough to become black holes—called "strange stars."

[3] A magical, unbreakable gun that you could hold without your arm being torn off. Don't worry, that part comes later!

fell, spraying in all directions. As it went in, you'd feel the ground shake, and it would leave a jumbled, fractured crater with no entry hole.

The bullet would fall straight through the Earth's crust. On the surface, the vibration would quickly die down. But far below, the bullet would be crushing and vaporizing the mantle in front of it as it fell. It would blast the material out of the way with powerful shockwaves, leaving a trail of superhot plasma behind it. This would be something never before seen in the history of the universe: an underground shooting star.

Eventually, the bullet would come to rest, lodged in the nickel-iron core at the center of the Earth. The energy delivered to the Earth would be massive on a human scale, but the planet would barely notice. The bullet's gravity would affect only the rock within a few dozen feet of it; while it's heavy enough to fall through the crust, its gravity alone wouldn't be strong enough to crush the rock very much.

The hole would close up, leaving the bullet forever out of anyone's reach.[4] Eventually, the Earth would be consumed by the aging, swollen Sun, and the bullet would reach its final resting place at the Sun's core.

The Sun isn't dense enough to become a neutron star itself. After it swallows the Earth, it will instead go through some phases of expansion and collapse, and will eventually settle down, leaving behind a small white dwarf star with the bullet still lodged in the center. Someday, far in the future—when the universe is thousands of times older than it is today—that white dwarf will cool and fade to black.

---

4 . . . unless Kyp Durron uses the Force to drag it back up.

That answers the question of what would happen if the bullet were fired into the Earth. But what if we could keep it near the surface?

Set the bullet on a sturdy pedestal

First, we'd need a magical infinitely strong pedestal to put the bullet on, which would need to sit on a similarly strong platform large enough to spread the weight out. Otherwise, the whole thing would sink into the ground.

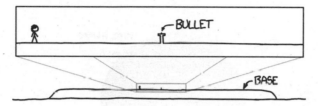

A base about the size of a city block would be strong enough to keep it above-ground for at least a few days, probably much more. After all, the Empire State Building—which weighs as much as our bullet—rests on a similar platform, and it's more than a few days old[citation needed] and hasn't disappeared into the ground.[citation needed]

The bullet wouldn't vacuum up the atmosphere. It would definitely compress the air around it and warm it up a little, but surprisingly, not really enough to notice.

Can I touch it?

Let's imagine what would happen if you tried.

The gravity from this thing is strong. But it's not *that* strong.

Imagine you're standing 10 meters away. At this distance, you feel a very slight tug in the direction of the pedestal. Your brain—not accustomed to nonuniform gravities—thinks you're standing on a gentle slope.

Do not put on roller skates.

This perceived slope gets steeper as you walk toward the pedestal, as if the ground were tipping forward.

When you get within a few meters, you have a hard time not sliding forward. However, if you got a good grip on something—a handle or a signpost—you can get pretty close.

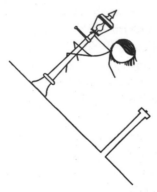

Los Alamos physicists might call this "tickling the dragon's tail."

But I wanna touch it!

To get close enough to touch it, you would need a *very* good grip on something. Really, you'd need to do this in a full-body support harness, or at the very least a neck brace; if you get within reach, your head will weigh as much as a small child, and your blood won't know which way to flow. However, if you're a fighter pilot who's used to gee forces, you might be able to pull it off.

From this angle, the blood is rushing to your head, but you'd still be able to breathe.

As you stretch out your arm, the pull gets a *lot* stronger; 20 centimeters (about 8 inches) is the point of no return—as your fingertips cross that line, your arm becomes too heavy to pull back. (If you do a lot of one-handed pull-ups, you might be able to go a little closer.)

Once you get within a few inches, the force on your fingers is overwhelming, and they're yanked forward—with or without you—and your fingertips actually touch the bullet (probably dislocating your fingers and shoulder).

When your fingertip actually comes in contact with the bullet, the pressure in your fingertips becomes too strong, and your blood breaks through the skin.

In *Firefly*, River Tam famously commented that "the human body can be drained of blood in 8.6 seconds given adequate vacuuming systems."

By touching the bullet, you've just created an adequate vacuuming system.

Your body is restrained by a harness, and your arm remains attached to your body—flesh is surprisingly strong—but blood pours from your fingertip much faster than ordinarily possible. River's "8.6 seconds" might be an underestimate.

Then things get weird.

The blood wraps around the bullet, forming a growing dark red sphere whose surface hums and vibrates with ripples moving too fast to see.

But wait

There's a fact that now becomes becomes important:

You *float* on blood.

As the blood sphere grows, the force on your shoulder weakens . . . because the parts of your fingertips below the surface of the blood are buoyant! Blood is denser than flesh, and half the weight on your arm was coming from the last two knuckles of your fingers. When the blood is a few centimeters deep, the load gets considerably lighter.

If you could wait for the sphere of blood to get 20 centimeters deep—and if your shoulder were intact—you might even be able to pull your arm away.

Problem: That would take five times as much blood as you have in your body.

It looks like you're not going to make it.

Let's rewind.

How to touch a neutron bullet: salt, water, and vodka

You can touch the bullet and survive . . . but you need to surround it with water.

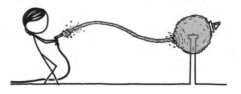

DO try this at home, and send me videos.

If you want to be really clever, you can dangle the end of the hose in the water and let the bullet's gravity do the siphoning for you.

To touch the bullet, pour water onto the pedestal until it's a meter or 2 deep on the side of the bullet. It will form a shape like one of these:

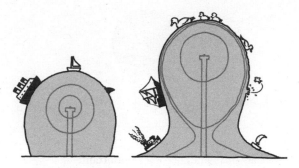

If those boats sink, you're not salvaging them.

Now, dip your head and arm in.

Thanks to the water, you're able to wave your hand around the bullet without any difficulty! The bullet is pulling you toward it, but it's pulling the water just as hard. Water (like meat) is virtually incompressible, even at these pressures, so nothing critical gets crushed.[5]

However, you may not quite be able to touch the bullet. When your fingers get a few millimeters away, the powerful gravity means that buoyancy plays a gigantic role. If your hand is slightly less dense than the water, it won't be able to penetrate that last millimeter. If it's slightly more dense, it will be sucked down.

This is where the vodka and salt come in. If you find the bullet tugging on your fingertips as you reach in, it means your fingers aren't buoyant enough. Mix in some salt to make the water denser. If you find your fingertips sliding on an invisible surface at the edge of the bullet, make the water less dense by adding vodka.

If you got the balance just right, you could touch the bullet and live to tell about it.

Maybe.

Alternative plan

Sound too risky to you? No problem. This whole plan—the bullet, the water, the salt, the vodka—doubles as instructions for making the most difficult mixed drink in the history of beverages: the **Neutron Star.**

---

5 When you pull your arm out, watch for symptoms of decompression sickness due to nitrogen bubbles in the blood vessels in your hand.

So grab a straw and take a drink.

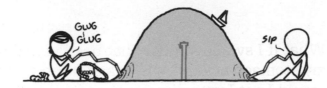

. . . and remember: If someone drops a cherry into your Neutron Star, and it sinks to the bottom, don't try to fish it out. It's gone.

WEIRD (AND WORRYING) QUESTIONS FROM THE WHAT IF? INBOX, #12

Q. What if I swallow a tick that has Lyme disease? Will my stomach acid kill the tick and the borreliosis, or would I get Lyme disease from the inside out?

—Christopher Vogel

JUST TO BE SAFE, YOU SHOULD SWALLOW SOMETHING TO KILL THE TICK, LIKE *SOLENOPSIS GERMINATA* (TROPICAL FIRE ANT). THEN, SWALLOW A *PSEUDACTEON CURVATUS* FLY TO KILL THE ANT. NEXT, FIND A SPIDER...

Q. Assuming a relatively uniform resonant frequency in a passenger jet, how many cats, meowing at what resonant frequency of said jet, would be required to "bring it down"?

—Brittany

HELLO, FAA? IS THERE A "BRITTANY" ON THE NO FLY LIST? ...YES, WITH CATS. THAT SOUNDS LIKE HER. OK, JUST MAKING SURE YOU WERE AWARE.

Q. What if a Richter magnitude 15 earthquake were to hit America at, let's say, New York City? What about a Richter 20? 25?

—Alec Farid

A. THE RICHTER SCALE, WHICH has technically been replaced by the "moment magnitude"[1] scale, measures the energy released by an earthquake. It's an open-ended scale, but since we usually hear about earthquakes with ratings from 3 to 9, a lot of people probably think of 10 as the top and 1 as the bottom.

In fact, 10 *isn't* the top of the scale, but it might as well be. A magnitude 9 earthquake already measurably alters the rotation of the Earth; the two magnitude 9+ earthquakes this century both altered the length of the day by a tiny fraction of a second.

A magnitude 15 earthquake would involve the release of almost 10^{32} joules of energy, which is roughly the gravitational binding energy of the Earth. To put it another way, the Death Star caused a magnitude 15 earthquake on Alderaan.

---

1   Similarly, the F-scale (Fujita scale) has been replaced by the EF-scale ("Enhanced Fujita"). Sometimes, a unit of measure is made obsolete because it is terrible—for example, "kips" (1000 pounds-force), "kcfs" (thousands of cubic feet per second), and "degrees Rankine" (degrees Fahrenheit above absolute zero). (I have had to read technical papers written in each of those units.) Other times, you get the sense that scientists just want something to correct people about.

You could in theory have a more powerful earthquake on Earth, but in practice all it would mean is that the expanding cloud of debris would be hotter.

The Sun, with its higher gravitational binding energy, could have a magnitude 20 quake (although it would certainly trigger some kind of a catastrophic nova). The most powerful quakes in the known universe, which occur in the material in a superheavy neutron star, are about this magnitude. This is about the energy release you would get if you packed the entire volume of the Earth with hydrogen bombs and detonated them all at once.

We spend a lot of time talking about things that are large and violent. But what about the *bottom* end of the scale? Is there such a thing as a magnitude 0 earthquake?

Yes! In fact, the scale goes all the way down *past* zero. Let's take a look at some low-magnitude "earthquakes," with a description of what they would be like if they hit your house.

Magnitude 0

The Dallas Cowboys running at full tilt into the side of your neighbor's garage

Magnitude -1

A single football player running into a tree in your yard

Magnitude -2

A cat falling off a dresser

Magnitude -3

A cat knocking your cell phone off your nightstand

Magnitude -4

A penny falling off a dog

Magnitude -5

A key press on an IBM model M keyboard

Magnitude -6

A key press on a lightweight keyboard

Magnitude -7

A single feather fluttering to the ground

Magnitude -8

A grain of fine sand falling onto the pile at the bottom of a tiny hourglass

. . . and let's jump all the way down to

Magnitude -15

A drifting mote of dust coming to rest on a table

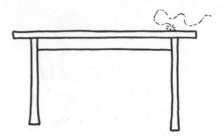

Sometimes it's nice *not* to destroy the world for a change.

ACKNOWLEDGMENTS

A bunch of people helped me make this book you're looking at.

Thank you to my editor, Courtney Young, for being an xkcd reader from the beginning and seeing this book through to the end. Thank you to the various terrific people at HMH who made everything work. Thank you to Seth Fishman and the Gernert folks for being patient and tireless.

Thank you to Christina Gleason for making this book look like a book, even when it meant deciphering my scribbled notes about asteroids at three in the morning. Thank you to the various experts who helped me answer questions, including Reuven Lazarus and Ellen McManis (radiation), Alice Kaanta (genes), Derek Lowe (chemicals), Nicole Gugliucci (telescopes), Ian Mackay (viruses), and Sarah Gillespie (bullets). Thank you to davean, who made this all happen but hates attention and will probably complain about being mentioned here.

Thank you to the IRC crowd for their comments and corrections, and to Finn, Ellen, Ada, and Ricky for sifting through the flood of submitted questions and filtering out the ones about Goku. Thank you to Goku for apparently being an anime character with infinite strength, and thus provoking hundreds of What If questions, even though I refused to watch *Dragon Ball Z* in order to answer them.

Thank you to my family for teaching me to answer absurd questions by spending so many years patiently answering mine. Thank you to my father for teaching me about measurement and my mother for teaching me about patterns. And thank you to my wife, for teaching me how to be tough, teaching me how to be brave, and teaching me about birds.

REFERENCES

Global Windstorm

Merlis, Timothy M., and Tapio Schneider, "Atmospheric dynamics of Earth-like tidally locked aquaplanets," *Journal of Advances in Modeling Earth Systems 2* (December 2010); DOI:10.3894/JAMES.2010.2.13.

"What Happens Underwater During a Hurricane?" *http://www.rsmas.miami.edu/blog/2012/10/22 /what-happens-underwater-during-a-hurricane*

Spent Fuel Pool

"Behavior of spent nuclear fuel in water pool storage," *http://www.osti.gov/energycitations/servlets/purl /7284014-xaMiio/7284014.pdf*

"Unplanned Exposure During Diving in the Spent Fuel Pool," *http://www.isoe-network.net/index.php/publications -mainmenu-88/isoe-news/doc_download/1756-ritter2011ppt.html*

Laser Pointer

GOOD, "Mapping the World's Population by Latitude, Longitude," *http://www.good.is/posts /mapping-the-world-s-population-by-latitude-longitude*

http://www.wickedlasers.com/arctic

Periodic Wall of the Elements

Table on page 9 (publication page 15, pdf page 15) in *http://www.epa.gov/opptintr/aegl/pubs /arsenictrioxide_po1_1sddelete.pdf*

Everybody Jump

Dot Physics, "What if everyone jumped?" *http://scienceblogs.com/dotphysics/2010/08/26 /what-if-everyone-jumped/*

Straight Dope, "If everyone in China jumped off chairs at once, would the earth be thrown out of its orbit?" *http://www.straightdope.com/columns/read/142 /if-all-chinese-jumped-at-once-would-cataclysm-result*

A Mole of Moles

Discover, "How many habitable planets are there in the galaxy?" *http://blogs.discovermagazine.com/badastronomy/2010/10/29 /how-many-habitable-planets-are-there-in-the-galaxy*

Hair Dryer

"Determination of Skin Burn Temperature Limits for Insulative Coatings Used for Personnel Protection," *http://www.mascoat.com/assets/files /Insulative_Coating_Evaluation_NACE.pdf*

"The Nuclear Potato Cannon Part 2," *http://nfttu.blogspot.com/2006/01 /nuclear-potato-cannon-part-2.html*

The Last Human Light

"Wind Turbine Lubrication and Maintenance: Protecting Investments in Renewable Energy," *http://www.renewableenergyworld.com/rea/news/article/2013/05 /wind-turbine-lubrication-and-maintenance-protecting -investments-in-renewable-energy*

McComas, D.J., J.P. Carrico, B. Hautamaki, M. Intelisano, R. Lebois, M. Loucks, L. Policastri, M. Reno, J. Scherrer, N.A. Schwadron, M. Tapley, and R. Tyler, "A new class of long–term stable lunar resonance orbits: Space weather applications and the Interstellar Boundary Explorer," *Space Weather*, 9, S11002, doi: 10.1029/2011SW000704, 2011.

Swift, G.M., et al. "In-flight annealing of displacement damage in GaAs LEDs: A Galileo story," *IEEE Transactions on Nuclear Science*, Vol. 50, Issue 6 (2003).

"Geothermal Binary Plant Operation and Maintenance Systems with Svartsengi Power Plant as a Case Study," *http://www.os.is/gogn/unu-gtp-report/UNU-GTP-2002-15.pdf*

Machine-Gun Jetpack

"Lecture L14 - Variable Mass Systems: The Rocket Equation" *http://ocw.mit.edu/courses/aeronautics-and-astronautics /16-07-dynamics-fall-2009/lecture-notes /MIT16_07F09_Lec14.pdf*

"[2.4] Attack Flogger in Service," *http://www.airvectors.net/avmig23_2.html#m4*

Rising Steadily

Otis: "About Elevators," *http://www.otisworldwide.com/pdf/AboutElevators.pdf*

National Weather Service: "Wind Chill Temperature Index," *http://www.nws.noaa.gov/om/windchill/images /wind-chill-brochure.pdf*

"Prediction of Survival Time in Cold Air"—see page 24 for the relevant tables,
http://cradpdf.drdc-rddc.gc.ca/PDFS/zba6/p144967.pdf

Linda D. Pendleton, "When Humans Fly High: What Pilots Should Know About High-Altitude Physiology, Hypoxia, and Rapid Decompression." *http://www.avweb.com /news/aeromed/181893-1.html*

Short-Answer Section

"Currency in Circulation: Volume,"
http://www.federalreserve.gov/paymentsystems /coin_currcircvolume.htm

NOAA, "Subject: C5c, Why don't we try to destroy tropical cyclones by nuking them?"
http://www.aoml.noaa.gov/hrd/tcfaq/C5c.html

NASA, "Stagnation Temperature,"
http://www.grc.nasa.gov/WWW/BGH/stagtmp.html

Lightning

"Lightning Captured @ 7,207 Fps,"
http://www.youtube.com/watch?v=BxQt8ivUGWQ

NOVA, "Lightning: Expert Q&A,"
http://www.pbs.org/wgbh/nova/earth/dwyer-lightning.html

JGR, "Computation of the diameter of a lightning return stroke"
http://onlinelibrary.wiley.com/doi/10.1029/JB073i006p01889 /abstract

Human Computer

"Moore's Law at 40,"
http://www.ece.ucsb.edu/~strukov/ece15bSpring2011/others /MooresLawat40.pdf

Little Planet

For another take on *The Little Prince*, scroll down to the last section of this wonderful piece by Mallory Ortberg,
http://the-toast.net/2013/08/02/texts-from-peter-pan-et-al/

Rugescu, Radu D., and Daniele Mortari, "Ultra Long Orbital Tethers Behave Highly Non-Keplerian and Unstable," *WSEAS Transactions on Mathematics*, Vol. 7, No. 3, March 2008, pp. 87–94,
http://www.academia.edu/3453325/Ultra_Long_Orbital_Tethers _Behave_Highly_Non-Keplerian_and_Unstable

Steak Drop

"Falling Faster than the Speed of Sound,"
http://blog.wolfram.com/2012/10/24 /falling-faster-than-the-speed-of-sound

"Stagnation Temperature: Real Gas Effects,"
http://www.grc.nasa.gov/WWW/BGH/stagtmp.html

"Predictions of Aerodynamic Heating on Tactical Missile Domes,"
http://www.dtic.mil/cgi-bin/GetTRDoc?AD=ADA073217

"Calculation of Reentry-Vehicle Temperature History,"
http://www.dtic.mil/dtic/tr/fulltext/u2/a231552.pdf

"Back in the Saddle,"
http://www.ejectionsite.com/insaddle/insaddle.htm

"How to Cook Pittsburgh-Style Steaks,"
http://www.livestrong.com/article /436635-how-to-cook-pittsburgh-style-steaks

Hockey Puck

"KHL's Alexander Ryazantsev sets new 'world record' for hardest shot at 114 mph,"
http://sports.yahoo.com/blogs/nhl-puck-daddy/khl-alexander -ryazantsev-sets-world-record-hardest-shot-174131642.html

"Superconducting Magnets for Maglifter Launch Assist Sleds,"
http://www.psfc.mit.edu/~radovinsky/papers/32.pdf

"Two-Stage Light Gas Guns,"
http://www.nasa.gov/centers/wstf/laboratories/hypervelocity /gasguns.html

"Hockey Video: Goalies, Hits, Goals, and Fights,"
http://www.youtube.com/watch?v=fWj6--Cf9QA

Common Cold

P. Stride, "The St. Kilda boat cough under the microscope," *The Journal— Royal College of Physicians of Edinburgh*, 2008; 38:272–9.

L. Kaiser, J. D. Aubert, et al., "Chronic Rhinoviral Infection in Lung Transplant Recipients," *American Journal of Respiratory and Critical Care Medicine*, Vol. 174; pp. 1392–1399, 2006, 10.1164/rccm.200604-489OC

Oliver, B. G. G., S. Lim, P. Wark, V. Laza-Stanca, N. King, J. L. Black, J. K. Burgess, M. Roth, and S. L. Johnston, "Rhinovirus Exposure Impairs Immune Responses To Bacterial Products In Human Alveolar Macrophages," *Thorax* 63, no. 6 (2008): 519–525.

Glass Half Empty

"Shatter beer bottles: Bare-handed bottle smash,"
http://www.youtube.com/watch?v=77gWkloZUC8

Alien Astronomers

The Hitchhiker's Guide to the Galaxy,
http://www.goodreads.com/book/show /11.The_Hitchhiker_s_Guide_to_the_Galaxy

"A Failure of Serendipity: The Square Kilometre Array will struggle to eavesdrop on Human-like ETI,"
http://arxiv.org/PS_cache/arxiv/pdf/1007/1007.0850v1.pdf

"Eavesdropping on Radio Broadcasts from Galactic Civilizations with Upcoming Observatories for Redshifted 21cm Radiation,"
http://arxiv.org/pdf/astro-ph/0610377v2.pdf

"The Earth as a Distant Planet a Rosetta Stone for the Search of Earth-Like Worlds,"
http://www.worldcat.org/title/earth-as-a-distant-planet -a-rosetta-stone-for-the-search-of-earth-like-worlds /oclc/643269627

"SETI on the SKA,"
http://www.astrobio.net/exclusive/4847/seti-on-the-ska

Gemini Planet Imager,
http://planetimager.org/

No More DNA

Enjalbert, Françoise, Sylvie Rapior, Janine Nouguier-Soulé, Sophie Guillon, Noël Amouroux, and Claudine Cabot, "Treatment of Amatoxin Poisoning: 20-Year Retrospective Analysis." *Clinical Toxicology* 40, no. 6 (2002): 715–757.
http://toxicology.ws/LLSAArticles/Treatment%20of %20Amatoxin%20Poisoning-20%20year%20retrospective %20analysis%20(J%20Toxicol%20Clin%20Toxicol%202002).pdf

Richard Eshelman, "I nearly died after eating wild mushrooms," *The Guardian* (2010), *http://www.theguardian.com/lifeandstyle/2010/nov/13 /nearly-died-eating-wild-mushrooms*

"Amatoxin: A review," *http://www.omicsgroup.org/journals/2165-7548/2165-7548-2-110 .php?aid=5258*

Interplanetary Cessna

"The Martian Chronicles," *http://www.x-plane.com/adventures/mars.html*

"Aerial Regional-Scale Environmental Survey of Mars," *http://marsairplane.larc.nasa.gov/*

"Panoramic Views and Landscape Mosaics of Titan Stitched from Huygens Raw Images," *http://www.beugungsbild.de/huygens/huygens.html*

"New images from Titan," *http://www.esa.int/Our_Activities/Space_Science /Cassini-Huygens/New_images_from_Titan*

Yoda

Saturday Morning Breakfast Cereal, *http://www.smbc-comics.com/index .php?db=comics&id=2305#comic*

Youtube, "'Beethoven Virus'—Musical Tesla Coils," *http://www.youtube.com/watch?v=uNJjnx-GdlE*

"Beast." The 15Kw 7' tall DR (DRSSTC 5), *http://www.goodchildengineering.com/tesla-coils /drsstc-5-10kw-monster*

Falling with Helium

De Haven, H., "Mechanical analysis of survival in falls from heights of fifty to one hundred and fifty feet," *Injury Prevention*, 6(1):62-b-68, *http://injuryprevention.bmj.com/content/6/1/62.3.long*

"Armchair Airman Says Flight Fulfilled His Lifelong Dream," *New York Times*, July 4, 1982, *http://www.nytimes.com/1982/07/04/us/armchair-airman-says -flight-fulfilled-his-lifelong-dream.html?pagewanted=all*

Jason Martinez, "Falling Faster than the Speed of Sound," Wolfram Blog, October 24, 2012, *http://blog.wolfram.com/2012/10/24 /falling-faster-than-the-speed-of-sound*

Everybody Out

George Dyson, *Project Orion: The True Story of the Atomic Spaceship* (New York: Henry Holt and Company, 2002)

Self-Fertilization

"Sperm Cells Created From Human Bone Marrow," *http://www.sciencedaily.com/releases/2007/04/070412211409.htm*

Nayernia, Karim, Tom Strachan, Majlinda Lako, Jae Ho Lee, Xin Zhang, Alison Murdoch, John Parrington, Miodrag Stojkovic, David Elliott, Wolfgang Engel, Manyu Li, Mary Herbert, and Lyle Armstrong, "RETRACTION - In Vitro Derivation Of Human Sperm From Embryonic Stem Cells," *Stem Cells and Development* (2009): 0908w75909069.

"Can sperm really be created in a laboratory?" *http://www.theguardian.com/lifeandstyle/2009/jul/09 /sperm-laboratory-men*

This is discussed more deeply in F. M. Lancaster's monograph Genetic and Quantitative Aspects of Genealogy at *http://www.genetic-genealogy.co.uk/Toc115570144.html*.

High Throw

"A Prehistory of Throwing Things," *http://ecodevoevo.blogspot.com/2009/10 /prehistory-of-throwing-things.html*

"Chapter 9. Stone tools and the evolution of hominin and human cognition," *http://www.academia.edu/235788/Chapter_9._Stone_tools_and _the_evolution_of_hominin_and_human_cognition*

"The unitary hypothesis: A common neural circuitry for novel manipulations, language, plan-ahead, and throwing?" *http://www.williamcalvin.com/1990s/1993Unitary.htm*

"Evolution of the human hand: The role of throwing and clubbing," *http://www.ncbi.nlm.nih.gov/pmc/articles/PMC1571064*

"Errors in the control of joint rotations associated with inaccuracies in overarm throws," *http://jn.physiology.org/content/75/3/1013.abstract*

"Speed of Nerve Impulses," *http://hypertextbook.com/facts/2002/DavidParizh.shtml*

"Farthest Distance to Throw a Golf Ball," *http://recordsetter.com/world-record /world-record-for-throwing-golf-ball/7349#contentsection*

Lethal Neutrinos

Karam, P. Andrew. "Gamma and Neutrino Radiation Dose from Gamma Ray Bursts and Nearby Supernovae," *Health Physics* 82, no. 4 (2002): 491–99.

Speed Bump

"Speed bump-induced spinal column injury," *http://akademikpersonel.duzce.edu.tr/hayatikandis/sci /hayatikandis12.01.2012_08.54.59sci.pdf*

"Speed hump spine fractures: Injury mechanism and case series," *http://www.ncbi.nlm.nih.gov/pubmed/21150664*

"The 2nd American Conference on Human Vibration," *http://www.cdc.gov/niosh/mining/UserFiles/works /pdfs/2009-145.pdf*

"Speed bump in Dubai + flying Gallardo," *http://www.youtube.com/watch?v=Vg79_mM7CNY*

Parker, Barry R., "Aerodynamic Design," *The Isaac Newton School of Driving: Physics and your car.* Baltimore, MD: Johns Hopkins University Press, 2003, 155.

The Myth of the 200-mph "Lift-Off Speed," *http://www.buildingspeed.org/blog/2012/06 /the-myth-of-the-200-mph-lift-off-speed/*

"Mercedes CLR-GTR Le Mans Flip," *http://www.youtube.com/watch?v=rQbgSe9S54I*

National Highway Transportation NHTSA, Summary of State Speed Laws, 2007

FedEx Bandwith

"FedEx still faster than the Internet," *http://royal.pingdom.com/2007/04/11 /fedex-still-faster-than-the-internet*

"Cisco Visual Networking Index: Forecast and Methodology, 2012–2017,"
http://www.cisco.com/en/US/solutions/collateral/ns341/ns525/ ns537/ns705/ns827/white_paper_c11-481360_ns827_Networking _Solutions_White_Paper.html

"Intel® Solid-State Drive 520 Series,"
http://download.intel.com/newsroom/kits/ssd/pdfs /intel_ssd_520_product_spec_325968.pdf

"Trinity test press releases (May 1945),"
http://blog.nuclearsecrecy.com/2011/11/10/weekly-document-01

"NEC and Corning achieve petabit optical transmission,"
http://optics.org/news/4/1/29

Free Fall

"Super Mario Bros.—Speedrun level 1 - 1 [370],"
http://www.youtube.com/watch?v=DGQGvAwqpbE

"Sprint ring cycle,"
http://www1.sprintpcs.com/support /HelpCenter.jsp?FOLDER%3C%3Efolder_id=1531979#4

"Glide data,"
http://www.dropzone.com/cgi-bin/forum /gforum.cgi?post=577711#577711

"Jump. Fly. Land.," *Air & Space,*
http://www.airspacemag.com/flight-today/Jump-Fly-Land.html

Prof. Dr. Herrligkoffer, "The East Pillar of Nanga Parbat," *The Alpine Journal* (1984).

The Guestroom, "Dr. Glenn Singleman and Heather Swan,"
http://www.abc.net.au/local/audio/2010/08/24/2991588.htm

"Highest BASE jump: Valery Rozov breaks Guinness world record,"
http://www.worldrecordacademy.com/sports /highest_BASE_jump_Valery_Rozov_breaks _Guinness_world_record_213415.html

Dean Potter, "Above It All,"
http://www.tonywingsuits.com/deanpotter.html

Sparta

According to a random stranger on the Internet, Andy Lubienski, "The Longbow,"
http://www.pomian.demon.co.uk/longbow.htm

Drain the Oceans

Extrapolated from the maximum pressure tolerable by icebreaker ship hull plates: *http://www.iacs.org.uk/document/public/ Publications /Unified_requirements/PDF/UR_I_pdf410.pdf*

"An experimental study of critical submergence to avoid free-surface vortices at vertical intakes,"
http://www.leg.state.mn.us/docs/pre2003/other/840235.pdf

Drain the Oceans: Part II

Donald Rapp, "Accessible Water on Mars," JPL D-31343-Rev.7,
http://spaceclimate.net/Mars.Water.7.06R.pdf

D. L. Santiago et al., "Mars climate and outflow events,"
http://spacescience.arc.nasa.gov

D. L. Santiago et al., "Cloud formation and water transport on Mars after major outflow events," 43rd Planetary Science Conference (2012).

Maggie Fox, "Mars May Not Have Been Warm or Wet,"
http://rense.com/general32/marsmaynothave.htm

Twitter

The Story of Mankind,
http://books.google.com /books?id=RskHAAAAIAAJ&pg=PA1#v=onepage&q&f=false

"Counting Characters,"
https://dev.twitter.com/docs/counting-characters

"A Mathematical Theory of Communication,"
http://cm.bell-labs.com/cm/ms/what/shannonday /shannon1948.pdf

Lego Bridge

"How tall can a Lego tower get?"
http://www.bbc.co.uk/news/magazine-20578627

"Investigation Into the Strength of Lego Technic Beams and Pin Connections,"
http://eprints.usq.edu.au/20528/1/Lostroh_LegoTesting_2012.pdf

"Total value of property in London soars to £1.35trn,"
http://www.standard.co.uk/business/business-news /total-value-of-property-in-london-soars-to-135trn-8779991.html

Random Sneeze Call

Cari Nierenberg, "The Perils of Sneezing, ABC News," Dec. 22, 2008.
http://abcnews.go.com/Health/ColdandFluNews /story?id=6479792&page=1

Bischoff Werner E., Michelle L. Wallis, Brian K. Tucker, Beth A. Reboussin, Michael A. Pfaller, Frederick G. Hayden, and Robert J. Sherertz, "'Gesundheit!' Sneezing, Common Colds, Allergies, and Staphylococcus aureus Dispersion," *J Infect Dis.* (2006), 194 (8): 1119–1126 doi:10.1086/507908

"Annual Rates of Lightning Fatalities by Country"
http://www.vaisala.com/Vaisala%20Documents/Scientific %20papers/Annual_rates_of_lightning_fatalities_by_country.pdf

Expanding Earth

"In conclusion, no statistically significant present expansion rate is detected by our study within the current measurement uncertainty of 0.2 mm yr-1."

Wu, X., X. Collilieux, Z. Altamimi, B. L. A. Vermeersen, R. S. Gross, and I. Fukumori (2011), "Accuracy of the International Terrestrial Reference Frame origin and Earth expansion, Geophys." Res. Lett., 38, L13304, doi:10.1029/2011GL047450,
http://repository.tudelft.nl/view/ir /uuid%3A72ed93c0-d13e-427c-8c5f-f013b737750e/

Lawrence Grybosky, "Thermal Expansion and Contraction,"
http://www.engr.psu.edu/ce/courses/ce584/concrete/library /cracking/thermalexpansioncontraction/thermalexpcontr.htm

Sasselov, Dimitar D., *The life of super-Earths: How the hunt for alien worlds and artificial cells will revolutionize life on our planet.* New York: Basic Books, 2012.

Franz, R.M. and P. C. Schutte, "Barometric hazards within the context of deep-level mining," *The Journal of The South African Institute of Mining and Metallurgy*

Plummer, H. C., "Note on the motion about an attracting centre of slowly increasing mass," *Monthly Notices of the Royal Astronomical Society*, Vol. 66, p. 83,
http://adsabs.harvard.edu/full/1906MNRAS..66...83P

Weightless Arrow

"*Hunting Arrow Selection Guide:* Chapter 5,"
http://www.huntersfriend.com/carbon_arrows /hunting_arrows_selection_guide_chapter_5.htm

"USA Archery Records, 2009,"
http://www.usaarcheryrecords.org/FlightPages/2009
/2009%20World%20Regular%20Flight%20Records.pdf

"Air flow around the point of an arrow,"
http://pip.sagepub.com/content/227/1/64.full.pdf

STS-124: KIBO, NASA,
http://www.nasa.gov/pdf/228145main_sts124_presskit2.pdf

Sunless Earth

"The 1859 Solar–Terrestrial Disturbance and the Current Limits of Extreme Space Weather Activity,"
http://www.leif.org/research
/1859%20Storm%20-%2Extreme%20Space%20Weather.pdf

"The extreme magnetic storm of 1–2 September 1859,"
http://trs-new.jpl.nasa.gov/dspace
/bitstream/2014/8787/1/02-1310.pdf

"Geomagnetic Storms,"
http://www.oecd.org/governance/risk/46891645.pdf

"Normalized Hurricane Damage in the United States: 1900–2005,"
http://sciencepolicy.colorado.edu/admin/publication_files
/resource-2476-2008.02.pdf

"A Satellite System for Avoiding Serial Sun-Transit Outages and Eclipses,"
http://www3.alcatel-lucent.com/bstj/vol49-1970/articles
/bstj49-8-1943.pdf

"Impacts of Federal-Aid Highway Investments Modeled by NBIAS,"
http://www.fhwa.dot.gov/policy/2010cpr/chap7.htm#9

"Time zones matter: The impact of distance and time zones on services trade,"
http://eeecon.uibk.ac.at/wopec2/repec/inn/wpaper/2012-14.pdf

"Baby Fact Sheet,"
http://www.ndhealth.gov/familyhealth/mch/babyfacts
/Sunburn.pdf

"The photic sneeze reflex as a risk factor to combat pilots,"
http://www.ncbi.nlm.nih.gov/pubmed/8108024

"Burned by wild parsnip,"
http://dnr.wi.gov/wnrmag/html/stories/1999/jun99/parsnip.htm

Updating a Printed Wikipedia

BrandNew: "Wikipedia as a Printed Book,"
http://www.brandnew.uk.com/wikipedia-as-a-printed-book/

ToolServer: Edit rate,
http://toolserver.org/~emijrp/wmcharts/wmchart0001.php

QualityLogic: Cost of Ink Per Page Analysis, June 2012,
http://www.qualitylogic.com/tuneup/uploads
/docfiles/QualityLogic-Cost-of-Ink-Per-Page-Analysis
_US_1-Jun-2012.pdf

Sunset on the British Empire

"Eddie Izzard - Do you have a flag?"
http://www.youtube.com/watch?v=uEx5G-GOS1k

"This Sceptred Isle: Empire.
A 90 part history of the British Empire,"
http://www.bbc.co.uk/radio4/history/empire/map

"A Guide to the British Overseas Territories,"
http://www.telegraph.co.uk/news/wikileaks-files
/london-wikileaks/8305236/A-GUIDE-TO-THE-BRITISH
-OVERSEAS-TERRITORIES.html

"Trouble in Paradise,"
http://www.vanityfair.com/culture/features/2008/01/
pitcairn200801

"Long History of Child Abuse Haunts Island 'Paradise,'"
http://www.npr.org/templates/story/story.php?storyId=103569364

"JavaScript Solar Eclipse Explorer,"
http://eclipse.gsfc.nasa.gov/JSEX/JSEX-index.html

Stirring Tea

"Brawn Mixer, Inc., Principles of Fluid Mixing (2003),"
http://www.craneengineering.net/products/mixers/documents
/craneEngineeringPrinciplesOfFluidMixing.pdf

"Cooling a cup of coffee with help of a spoon,"
http://physics.stackexchange.com/questions/5265
/cooling-a-cup-of-coffee-with-help-of-a-spoon/5510#5510

All the Lightning

"Introduction to Lightning Safety," National Weather Service, Wilmington, Ohio,
http://www.erh.noaa.gov/iln/lightning/2012
/lightningsafetyweek.php

Bürgesser Rodrigo E., Maria G. Nicora, and Eldo E. Ávila, "Characterization of the lightning activity of Relámpago del Catatumbo," Journal of Atmospheric and Solar-Terrestrial Physics (2011),
http://wwlln.net/publications/avila.Catatumbo2012.pdf

Loneliest Human

BBC Future interview with Al Wolden (April 2, 2013),
http://www.bbc.com/future/story/
20130401-the-loneliest-human-being/1

Raindrop

"SSMI/SSMIS/TMI-derived Total Precipitable Water-North Atlantic,"
http://tropic.ssec.wisc.edu/real-time/mimic-tpw/natl/main.html

"Structure of Florida Thunderstorms Using High-Altitude Aircraft Radiometer and Radar Observations," Journal of Applied Meteorology,
http://rsd.gsfc.nasa.gov/912/edop/misc/1736.pdf

SAT Guessing

Cooper, Mary Ann, MD., "Disability, Not Death Is the Main Problem with Lightning Injury,"
http://www.uic.edu/labs/lightninginjury/Disability.pdf

National Oceanic and Atmospheric Administration (NOAA), "2008 Lightning Fatalities,"
http://www.nws.noaa.gov/om/hazstats/light08.pdf

Neutron Bullet

"Influence of Small Arms Bullet Construction on Terminal Ballistics,"
http://bsrlab.gatech.edu/AUTODYN/papers/paper162.pdf

McCall, Benjamin, "Q & A: Neutron Star Densities," University of Illinois,
http://van.physics.illinois.edu/qa/listing.php?id=16748

$\times 6000$ ... Kcal/...

Ks...

...800...

$\dfrac{767 \text{ kg}}{1.583 \times 10^{9} \text{ w}}$

24 billion...

Flow =

$-V_0 =$

f